P9-AGP-445

REGIONAL ATLAS OF BONE DISEASE

DATE DUE

NO 7 '97			
DE 3 '97			
NO 7 '00			
AP 2 2 '05			
MAY 1 1 2005			

DEMCO 38-296

12

REGIONAL ATLAS OF BONE DISEASE
A Guide to Pathologic and Normal Variation in the Human Skeleton

By

ROBERT W. MANN

Department of Anthropology
National Museum of Natural History
Smithsonian Institution
Washington, D.C.

SEAN P. MURPHY

National Museum of Health and Medicine
Armed Forces Institute of Pathology
Washington, D.C.

Foreword by O' Brian C. Smith
Introduction by Donald J. Ortner

CHARLES C THOMAS • PUBLISHER
Springfield • Illinois • U.S.A.

Riverside Community College
Library
4800 Magnolia Avenue
Riverside, California 92506

RC 930.4 .M35 1990

Mann, Robert W., 1949-

Regional atlas of bone
   disease

hroughout the World by

CHARLES C THOMAS • PUBLISHER
2600 South First Street
Springfield, Illinois 62794-9265

This book is protected by copyright. No part of
it may be reproduced in any manner without
written permission from the publisher.

© *1990 by* CHARLES C THOMAS • PUBLISHER

ISBN 0-398-05675-7

Library of Congress Catalog Card Number: 90-10753

With THOMAS BOOKS *careful attention is given to all details of manufacturing
and design. It is the Publisher's desire to present books that are satisfactory as to their
physical qualities and artistic possibilities and appropriate for their particular use.*
THOMAS BOOKS *will be true to those laws of quality that assure a good name
and good will.*

Printed in the United States of America
SC-R-3

Library of Congress Cataloging-in-Publication Data

Mann, Robert W., 1949–
    Regional atlas of bone disease : a guide to pathologic and normal
variation in the human skeleton / by Robert W. Mann, Sean P. Murphy
; foreword by O'Brian C. Smith ; introduction by Donald J. Ortner.
        p. cm.
    Includes bibliographical references.
    ISBN 0-398-05675-7
    1. Bones—Diseases—Atlases.   2. Paleopathology—Atlases.
I. Murphy, Sean P.   II. Title.
    [DNLM: 1. Bone Diseases—atlases. 2. Paleopathology—atlases.
WE 17 M282r]
RC930.4.M35   1990
616.7′1′00222—dc20
DNLM/DLC
for Library of Congress   90-10753
                                                                      CIP

To Adele D. Mann and Robert P. Murphy

FOREWORD

"Art is I, Science is We"
Claude Bernard

Enthusiasm. The most motivating force in a student is enthusiasm. Many bring it with them, already on fire for their particular area of interest. Most though are infected with it by their instructors and fellow students as a sense of discovery, for advancement and competency develops. Enthusiasm dwarfs things petty to science; egos, attitudes, personal agendas, and the like. It creates an aura of academic purity, an environment without fear where "we" is paramount, and "I" becomes a measure of capacity, not importance. It is a time where we can be smart together and we can be dumb together without pride or fear.

Cultivating enthusiasm is one of the hardest tasks for an educator, especially in students just entering an area of study. Many disciplines have their own language, because it requires precisely defined concepts to advance the field. The introductory student needs to acquire some of this to be facile in developing his knowledge and thinking, but too much can intimidate and dampen enthusiasm. The educator, well-versed in terminology, needs to introduce his topic in the language of the layman in order to assure communication. This latter is not an easy task because precision of concept suffers.

It is a bold step then for any introductory text to be written especially for the entering student. Colleagues who have already achieved their knowledge-base can always be critical of the authors license and charge over-simplification; and in part will always be right. My reply is that I've rarely found reference books to have a well-thumbed appearance. If I have to choose between precision and enthusiasm for the new student, it will always be enthusiasm! If the fire gets stoked, the opportunity for full potential is achieved.

Let the above be enough to explain this book to my colleagues. I hope too, that they will learn some things from the authors, because I did. For you, the most important reader, the newest generation, I welcome you as colleagues and invite you to these pages. Read! Enjoy! Discover! Think!

O'BRIAN C. SMITH
Assistant Professor of Pathology;
University of Tennessee School of
Medicine, Knoxville
Deputy Chief Medical Examiner;
State of Tennessee

INTRODUCTION

In most areas of biomedical research, a carefully defined descriptive and classificatory methodology exists and constitutes a basic research tool. In paleopathology this is not the case and we are just beginning to confront some long-standing problems in description and classification that have limited its development. Because of these problems there exists a substantial literature in paleopathology that is of minimal value in clarifying many of the questions that must be addressed if paleopathology is to reach its full potential as a legitimate biomedical discipline.

We must, for example, be able to build a data base that will provide information on both the antiquity, geographical distribution and evolutionary trends of disease. We need data that will help to clarify the complex relationships that exist between the many factors that affect the human response to disease, including: (1) disease organisms, (2) environmental factors (e.g., air pollution) that affect health, (3) nutrition and (4) the immune response of a patient to disease.

The first step in the study of skeletal paleopathology should be the development of a method to describe the abnormal conditions encountered in archeological human skeletons. It is both surprising and frustrating that after 150 years of research in paleopathology so little effort has been expended to develop a careful and comprehensive descriptive terminology much less a taxonomy of what is seen in skeletal specimens. Much of how we describe pathological conditions in archeological skeletons is derivative of medical and particularly orthopedic nomenclature and classificatory systems. Both systems have evolved over the past 150 years which adds another complication in interpreting the older literature in paleopathology. The major problem, however, is not one of semantics. Rather it is that many of the lesions and their distribution patterns in archeological dry bone specimens bear minimal relationship to features that are central in clinical orthopedic practice. What is really needed is a nomenclature and classificatory system that integrates all of the pathological information that is apparent in skeletal paleopathological specimens. Such a system would necessarily include orthopedic terms and classification where the features were closely related to those used in a clinical setting. There are, however, occasional conditions in paleo-pathological cases that are not well known in clinical orthopedic practice and a precise classificatory system might demonstrate relationships that previously had not been understood.

In working with both professional colleagues and graduate students I have, for many years, emphasized the importance of first describing carefully what one sees in cases of skeletal paleopathology. Careful description is timeless and, if done well, forever gives future readers of reports the option of reinterpreting your conclusions (i. e., diagnoses). Demographic data, including age and sex, are important factors in interpreting descriptive information. However, the type and distribu-tion patterns of abnormal bone tissue addition or destruction are funda-mental to any effective descriptive analysis. Is, for example, the added bone poorly organized (this typically means rapid growth) or well orga-nized (usually slow growth)? Do destructive lesions have well-defined margins with evidence of well-organized bony repair (circumscribed and generally less aggressive) or poorly-defined margins (permeative and generally more aggressive)? These and other features are all critical elements in any interpretation of a paleopathological case of skeletal disease.

The location of lesions within the skeleton provides an important link with clinical experience but one needs to be cautious in making such associations. In dry-bone paleopathological cases one often sees lesions that would not be apparent in clinical x-ray films and are thus not well known in the medical descriptive and classificatory systems. Indeed this is one of the significant potential contributions that careful study of paleopathological cases can make to an understanding of the skeletal manifestations in orthopedic pathology. A pathology based on dry-bone conditions also means that some distribution patterns of abnormal tissue within a pathological skeleton will vary from patterns established on the basis of radiology in living patients.

Careful description is not easy and I do not wish to underestimate the difficulty of the process. Nevertheless, most people can, with relative ease, learn to recognize the essential features of bone reaction to disease. The first step is, of course, a thorough knowledge of the gross anatomy of normal bone at all ages from fetal through old age. Archeological skele-tal samples are a wonderful source of anatomical knowledge since the entire age spectrum is usually represented. Classification or diagnosis is a much more complicated matter and for many cases encountered by the student of paleopathology, years of experience and a comprehensive knowledge of orthopedic pathology will be necessary and, even so, may not be possible.

At this time the student of skeletal paleopathology does not have a well defined and widely recognized terminology to use in describing pathological conditions. I am optimistic that this will be developed in the near future. In the interim the best option is to use terms that are part of the general lexicon we all share. Bone addition, bone destruction, porous bone, and bone spurs are examples of terms that have wide recognition in many disciplines and I encourage their use. Jargon is one of the biggest barriers to effective communication that exists and should be eliminated or, at the very least, kept to a minimum. At some point, however, you will need to acquire a working knowledge of medical terminology if only to understand and interpret the existing literature on paleopathology and communicate with medically trained colleagues.

The Regional Atlas of Bone Disease is an attempt to assist the beginning skeletal paleopathologist to recognize some of the more common pathological conditions that may be encountered in dry bone specimens. The authors explicitly insist that their endeavor be viewed as an initial step in any classificatory process. This is wise counsel, given the complexity of doing so. One of the fundamental problems for any classificatory system is that the bone reaction to disease is limited. In view of this it is not surprising that a given pathological condition (i.e., osseous response) may be the result of any one of several pathological processes.

The reader should also be aware of the strengths and weaknesses of a regional approach to skeletal paleopathology. Archeological skeletal samples often do not have complete skeletons. This is particularly true of older museum collections where only the skull and mandible may have been recovered. However, even where an attempt was made to excavate the entire skeleton the result is usually only partially successful. In this context a regional review of pathological conditions may be the only one possible and is certainly helpful. It is also true that many pathological conditions occur in a single location in the skeleton (solitary or unifocal conditions). A regional focus is generally adequate for such lesions.

However, regional approach is less helpful in multifocal pathological conditions. In this type of skeletal paleopathology, the distribution pattern of abnormal bone is a critical element in classification and the user of a regional approach will need to reconstruct the overall pattern by carefully reviewing the information for each region of the skeleton. A review of the distribution pattern of abnormal bone is important for classification but also contributes to the general understanding of pathogenesis in orthopedic disorders.

Despite this cautionary note, the beginning skeletal paleopathologist should find the Regional Atlas a helpful starting place when he or she encounters a skeletal abnormality in archeological burials. Remember, however, first carefully and fully describe what you see including what is occurring and where it appears in the skeleton. An attempt at diagnosis can then be made with the assurance that others will at least have the option of reaching a different diagnostic conclusion on the basis of the descriptive information you have provided should that be appropriate. The authors' counsel to seek advice on diagnosis from specialists in skeletal disease is wise. Keep in mind, however, that very few medical specialists have experience with dry-bone specimens and are often as baffled by a pathological case as is the osteologist. The orthopedist does, however, have the advantage of knowing what most of the diagnostic options are and this is a very useful point of departure.

DONALD J. ORTNER
Chairman
Department of Anthropology
National Museum of Natural History
Smithsonian Institution
Washington, D.C.

ACKNOWLEDGMENTS

The authors would like to express their deep debt of gratitude to Drs. William M. Bass, Hugh E. Berryman, Bruce Bradtmiller, Mr. Henry W. Case, Ms. Lee Meadows, Mr. Paul S. Sledzik, Drs. Leslie E. Eisenberg, Richard L. Jantz, Marc A. Kelley, Keith A. Manchester, Marc S. Micozzi, Douglas W. Owsley, Charlotte A. Roberts, T. Dale Stewart, Douglas H. Ubelaker, and P. Willey. It was through the friendships, teachings, and professional guidance of these scientists that this book came to fruition.

All illustrations were done by Robert W. Mann except Figures 7, 36, 39, 42 and 92 (Elizabeth C. Lockett); Figures 40, 54, and 55 (Marcia D. Bakry); Figure 27 (Neil Fallon); and Figure 81 (Jennifer B. Clark). All drawings were based on anatomical specimens at the Smithsonian Institution, the University of Tennessee, Knoxville, and the National Museum of Health and Medicine of the Armed Forces Institute of Pathology, Washington, D.C. The authors would like to extend a special debt of gratitude to Dr. Donald J. Ortner for devoting precious hours from his duties as Departmental Chair to this enterprise. Dr. O'Brian C. Smith contributed substantially to the writing of Chapters IV and XI. Dr. Milton Jacobs provided invaluable editorial suggestions. The opinions expressed in the Regional Atlas are the sole responsibility of the authors.

CONTENTS

REGIONAL ATLAS OF BONE DISEASE

Chapter I

USING THE REGIONAL ATLAS

The information contained in the Regional Atlas is based on paleo-pathological examination of over four thousand complete or nearly complete skeletons from archaeological sites in the Great Plains, Chile, Easter Island, northeastern United States, eight historic cemeteries (five from Louisiana and one each from Maryland, Virginia, and Washington, D.C.), War of 1812 and Civil War soldiers, and approximately one hundred forensic cases. For comparison, more contemporary skeletal samples from the Hamann-Todd (Cleveland, Ohio) and Terry Anatomical (Smithsonian Institution) collections were also studied.

The Regional Atlas approaches the recognition of disease according to the bone affected. The format of this handbook begins with a description of how to use the Regional Atlas (Chapter I), followed by a brief history of paleopathology (Chapter II). Chapter III gives step by step instructions on how the authors conduct a paleopathological analysis. Chapter IV briefly covers the mechanics of bone remodeling. The bulk of the Atlas is Chapter V which deals with specific diseases affecting each bone in the body beginning with the skull and progressing down the skeleton. Accompanying each lesion description is a statement of the relative frequency (e.g. uncommon to rare finding) or percentage that one might expect to find in most archaeological skeletal samples, especially American Indian groups. References cited within a sentence indicate that information was taken from these sources. References at the end of a paragraph (following a period) were included as additional sources if further information is desired by the reader.

Areas of rapidly growing bone in children are often mistaken for disease. Chapter VI, therefore, provides examples of a few bones of a child to alert the observer as to what is to be expected in a "healthy" child's skeleton. Fungal infections, due to their complex nature and similarity of lesions, are discussed in Chapter VII. The treponematoses (i.e., syphilis and allied conditions) are covered in Chapter VIII. Chapter IX briefly discusses tumors, perhaps the most difficult skeletal lesions to be diagnosed. Chapter X describes a hydatid cyst. This uncommon finding was included because of its unusual appearance and serves as an example of mineralization of soft tissue into bone by an invasive agent.

Peri-mortem and post-mortem fractures are discussed (Chapter XI) because of the difficulty in distinguishing the two conditions in archaeological specimens. Determining when a fracture occurred is important in establishing if the trauma was or was not related to the person's death. Chapter XII covers muscle attachment sites and is meant to serve as a quick reference to the musculoskeleton when ossification of a muscle, tendon, or ligament is suspected.

It should be remembered that no text can fully or even adequately cover every disease, anomaly, or normal anatomical variant present in the human skeleton; the present text is no exception. While some topics in the Regional Atlas are discussed in great detail, others are conspicuously brief owing to their extreme difficulty in differential diagnosis or rarity in most skeletal collections (e.g. tumors and fungal infections). One goal of the Regional Atlas was to include the findings and hypotheses of contemporary clinical practitioners (e.g. bone histologists, pathologists, rheumatologists, etc.) to supply the reader with a number of interpretations from which to choose. Such an approach also serves to educate the reader as to the complexity and controversy surrounding the identification, classification, and etiology of many bone diseases.

It is hoped that the experiences of the authors will make it possible for anyone with a sound knowledge of human osteology and skeletal morphology to conduct a basic *descriptive* paleopathological analysis of one or many skeletons. It should be noted, however, that the field of paleopathology is filled with ambiguities and subtleties. Putting this atlas to memory doesn't make one a paleopathologist; only knowledge, training, and above all, experience will qualify you for such a title. The Regional Atlas will, however, enable you to conduct your own analysis and, in questionable cases, alert you to seek the advice of an experienced paleopathologist, radiologist, or orthopaedist. The importance of a thorough descriptive analysis, however, cannot be overemphasized.

To use the Atlas, first become familiar with what and where lesions might be expected, locate and identify them in the text, and then refer to the excellent paleopathology and clinical texts by Brothwell and Sandison (1967), Cockburn and Cockburn (1980), Dieppe et al. (1986), Manchester

(1983), McCarty (1989), Moskowitz et al. (1984), Ortner and Putschar (1985), Resnick and Niwayama (1988), Steinbock (1976), Tyson and Dyer (1980), Wells (1964), Zimmerman and Kelley (1982), or other references in the text. It is hoped that the Regional Atlas will serve as a valuable companion to the existing literature on paleopathology.

Chapter II

A BRIEF HISTORY OF PALEOPATHOLOGY

Paleopathology is the study of disease in premodern anthropological or paleontological specimens. Research on this subject dates back nearly two hundred years and has evolved from an area devoted to the study of medical curiosities and exploration of the human body to an integrated discipline of medicine.

One of the earliest studies on bone diseases in dry specimens was made by Johann Esper who, in 1774, reported on a pathological femur of a cave bear from France. Since this earliest study, some scholars have drawn arbitrary temporal divisions marking periods where the primary focus of paleopathology has shifted. For example, before 1900 the focus was on traumatic lesions and syphilis, between 1900 and 1930 infectious disease. Today, the focus is comparative and multidisciplinary with ecology as a major component (Ubelaker 1982).

Some of the more notable early researchers in paleopathology include A. Hrdlicka, D. Brothwell, C.J. Hackett, E.A. Hooton, S. Jarcho, J. Jones, V. Moller-Christensen, R.L. Moodie, M. A. Ruffer, G.E. Smith, R. Virchow, F. Wood Jones, and C. Wells, to name a few. Each of these researchers made considerable contributions to the early study of bone disease in humans.

Although some early researchers studied large skeletal populations (e.g. Hooton 1930; Smith 1910), the population approach to disease frequencies was used by only a few, Wyman (1868) and Hrdlicka (1914) being notable examples. Because of the seemingly unlimited field of opportunity and discovery in paleopathology, much of the focus of bone diseases was on the more exciting and rare diseases and trauma such as tuberculosis, leprosy, syphilis, trephination, rickets, and the like. Skeletons exhibiting syphilitic-like lesions served to fuel many heated debates and spawned a number of hypotheses concerning its diagnosis in dry bone, as well as its time and place of origin.

While the "exciting" diseases were being debated by some of the

7

world's leading scholars, other researchers around the turn of the twentieth century focused on the less exotic skeletal indicators of health such as cribra orbitalia (Welcker 1888) and symmetric osteoporosis (Hrdlicka 1914). All of the early researchers had to rely on gross examination of the bones because radiographs, refinements in biochemistry, cytology (cell technology), and soft tissue pathology lay before them. Macroscopic examination and a descriptive analysis was the primary method of diagnosing bone disease when the "patient" was an archaeological skeleton with no clinical history. However, this did not stop some of the paleopathologists from offering their opinions and giving the disease a name (differential diagnosis), including ones that were incorrect. Although the cornerstone of contemporary paleopathological analysis continues to be the descriptive analysis, researchers now have at their disposal extremely sophisticated radiologic, immunologic, and microscopic techniques to aid in diagnosing skeletal lesions.

Contemporary paleopathology has carried on many of the traditions established by its forefathers. For example, much debate still continues on whether syphilitic-like lesions in archaeological bones represent venereal syphilis or one of the other treponematoses (e.g. yaws, endemic syphilis); the age-old debate of whether the Europeans spread syphilis to the New World also has yet to be resolved (Akrawi 1949; Bloch 1908; Brothwell 1970; Cockburn 1961; Hackett 1967; Holcomb 1930, 1935; Pusey 1915; Williams 1932) although innovative immunologic and multidisciplinary research may soon clarify this issue.

While the exotic diseases are still of keen interest to paleopathologists, an added emphasis has emerged focusing on the subtle bony changes that reflect physical stresses and activities of everyday life (e.g. mild indicators of osteoarthritis), nutrition, and diet (Angel 1976; Angel et al. 1987; Cook 1979; Cook and Buikstra 1979; Mann et al. 1987; Owsley et al. 1987; Schoeninger 1979; Sillen 1981).

Paleopathologists, through the cumulative studies of disease in the human skeleton, now have a better understanding of how many diseases developed and spread both geographically and temporally. The "mundane" skeletal indicators of physical stress, nutrition, and diet are now studied with the same enthusiasm as the exotic diseases. Only by including the full range of bone disease and testing hypotheses are we able to make valid biocultural conclusions on the general and specific health status of skeletal populations. As the late Dr. Larry Angel (1981) once wrote, "We are still all amateurs in paleopathology looking toward a bright cooperative future." For a more thorough history of paleopathology refer to Ubelaker (1982).

Chapter III

PREPARING FOR A
BASIC PALEOPATHOLOGICAL ANALYSIS

In preparing to examine a skeletal series it is important to first determine the focus of the study. It is simply not possible to gather every bit
of information on every skeleton. While one researcher might think it
imperative to take bone core samples for lead and other trace mineral
analyses, another may restrict his or her research to nondestructive
techniques. Prior written permission should always be obtained before
performing any destructive or invasive bone studies.

Before beginning the skeletal analysis you might first like to review, or
at least have at hand, some of the more comprehensive texts on human
osteology including Bass (1987), Steele and Bramblett (1988), Brothwell
(1981), Krogman and Iscan (1986), Shipman et al. (1985), Stewart (1979),
and Ubelaker (1989). As with any project, planning and organization are
the keys to a successful skeletal study.

The first order of business is to clarify the difference between a
"pathology" and a lesion or pathological condition. A "pathology" is not
synonymous with a lesion; pathology is the study of disease, whereas a
lesion is the response to disease or a wound (Thomas 1985). Although
many learned paleopathologists and clinicians use the words synony-
mously, the accurate term for a disease state in bone is either lesion,
wound, injury or pathological condition.

The following suggestions are offered as guidelines on how to prepare
for a paleopathological analysis. The technique can be applied to one or
one thousand skeletons and has proven itself to be comprehensive enough
to satisfy the needs of most researchers. Much of this technique was
developed by Drs. Douglas Owsley and Bruce Bradtmiller and has since
been successfully implemented on thousands of skeletons.

First lay out each skeleton on a separate tray in the correct anatomical
order. For example, place the ribs in proper order according to number
and side (e.g. left 1st, 2nd, 3rd, etc.). The vertebrae should also be laid

11

out beginning with the first cervical and ending with the fifth lumbar. The bones of the left side should be separated from those of the right as should the hands and feet, and so forth. This method allows pulling bones from various trays (or archaeological features) and comparing them for matches without commingling the bones.

When working with commingled skeletons (i.e., two or more skeletons mixed together) it is best to take standard 3 × 5 index cards, cut them in half, punch a hole in one corner, affix a rubber band, and attach a card to each bone with a burial or feature number. This method allows bones to be pulled from various trays (or archaeological features) and compared for matches without commingling the bones. Each element (bone) can later be put back on its original tray if no matches are made. Tagging the bones also allows them to be put aside for photographing and x-raying. After fragmentary skeletal elements are reconstructed (glued together) and inventoried, metrics, age, race, sex, and stature may be assigned.

Regarding the topic of reconstructing bones, brief mention of the pros and cons of particular adhesives is appropriate. The choice of adhesive will depend on a number of factors including whether the skeletons will be reburied, their stability, how long they will be available for study, will the bone be dated by C-14 method, and will the reconstruction be long-term (Pers. Comm. Greta Hansen 1989). Although Duco® Cement has been widely used for reconstructing bones, in time it will become dry, yellow, brittle, and separate at the glued "joints." The inherent instability of Duco Cement renders it unsuitable for long-term use. If the reconstruction is meant to be permanent, a better choice is Acryloid B-72 or B-48N (available from Rohm and Haas) or solvented polyvinyl acetate (PVA) (available from Union Carbide Corporation). For more information on reconstructing, preserving, and consolidating bone, consult an objects conservator or refer to Johansson (1987) or Sease (1987).

The next step in conducting a skeletal examination is to inspect each bone for pathological conditions. The most important technique in the analysis is examining each bone closely, using a standard fluorescent lamp. Hold the bone approximately 3–5 inches away from the lamp and inspect the surface inch by inch. Do not pick up a bone and quickly examine it for obvious pathological changes since many subtle lesions and, possibly, cutmarks from scalping or defleshing may be overlooked. It's a good idea to examine each bone by holding it in different positions to change the angle of the light as it hits the bone. This, too, may reveal a small lesion that was formerly hidden by shadows. If you find yourself staring at a questionable lesion, put the bone aside and return to it later. You might use this time to reference other paleopathology texts to help clarify your questions. Being consistent with your scoring criteria and judgment is critical.

If, after examining ten or fifteen skeletons, you find that you have been scoring a lesion as moderate in severity and feel that it should only be scored as slight, then you should modify the criteria for the trait or lesion in question. Obviously, it's easier to go back and change data sheets for ten or twenty skeletons than to realize this after you've analyzed one hundred or more. At such a late point in your examination, you might find that you've run out of time or are too frustrated to reexamine the skeletons. Be observant early in your analysis and note those conditions that are ambiguous in either severity or presence/absence. After the skeletons have been examined for pathological conditions, bones needing to be x-rayed or photographed should be set aside.

Note at the top of each inventory sheet and notes (separate pages for each individual) the burial number, age, sex, and any unusual features present in the bone, or associated with the particular burial (e.g. "projectile point in 3rd lumbar. Photo/x-ray"). Later this will save time relocating those skeletons requiring further study.

Finally, if you have access to a lap-top computer you can save many hours of deciphering, reexamining, and rewriting laboratory or field notes for each burial. This technique is especially important in a thorough *descriptive* paleopathology analysis. If you use the computer as a source of recording your methods, findings, and thoughts, a large part of the write up will already be done; this "rough draft" can then be edited and produces a very detailed report. (You'd be surprised how much information is forgotten after the analysis is completed.) The following is an example of how a skeleton might be described using a lap-top computer:

Skeleton 1. This is the poorly preserved skeleton of a young adult male, perhaps 20–25 years of age. Present are the skull, mandible (missing right ascending ramus), both femora, tibiae, left innominate, right foot, and left clavicle.

Age is based on examination of the pubic symphysis (billowy), auricular surface of the ilium (no microporosity), dental eruption, cranial suture closure (ectocranial incomplete), and medial epiphysis of the clavicle.

Sex is clearly male based on the morphology of the innominate (very narrow sciatic notch), robust skull (well-developed nuchal crest), and femur head diameter measuring 52 mm.

Pathological conditions: There is a roughly circular, healed depressed fracture in the frontal bone immediately above the right orbit. Fracture lines extend from the center of trauma and radiate into the superior orbital plate. There is no evidence of healing which suggests that this injury may have occurred at or near the time of death (peri-mortem). The inner vault of the skull was not affected (no fractures or depressed bone visible). The shape of the trauma site suggests that an oblong object struck this individual above the right eye. Examination of the remainder of the skeleton reveals no further evidence of trauma (many bones, however, are missing).

The distal right femur exhibits a small area of healed periostitis along the lateral surface. There is slight porosity (OA) in both temporal fossae (temporomandibular joints) and the right mandibular condyle is slightly flattened. The teeth show moderate attrition, two abscesses of the right mandibular first and second molars, and slight linear enamel hypoplasias.

Comments: The distal left radius exhibits a small area of green copper salts staining (about the size of a quarter) on its posterior surface (photo). The location of the stain suggests that a bracelet or other metal artifact was in contact with the left wrist at the time of burial. The field notes do not report any artifacts in association with this burial.

Chapter IV

FUNDAMENTALS OF BONE FORMATION AND REMODELING

The purpose of this chapter is to give a brief account of normal and pathological features of remodeled bone in the human skeleton. With a core knowledge of normal bone growth and remodeling, abnormal responses of bone can be appreciated by identifying any reactive changes in gross appearance. The full developmental sequence of the skeleton is left to the domain of the embryologist and pediatrician. Here basic facts are presented, sufficient to impart an understanding of bone formation and remodeling.

THE NECESSITY OF KNOWING

Recognizing a bone as abnormal raises questions about the cause or event (etiology) resulting in remodeling. Findings may be of forensic significance for the individual or reflect the culture of the population. Certain fracture patterns (parry fractures) may tell us of the warlike nature of the people or indicate the introduction of the horse (femoral/pelvic fractures). The ability to recognize infection (e.g. tuberculosis) helps us to understand the epidemiology of certain diseases and reflects hygiene or public health practices. Other changes may be characteristic of nutritional deficits (e.g. rickets), genetic predisposition to deformity (congenital) or tumor. But before any ability in recognizing disease can be acquired, a practical knowledge of normal must be obtained.

Functionally the skeleton supports the body, protects internal organs, and serves as attachment for the muscles and soft tissues. Bone is a living tissue consisting of 92 percent mineral or solids and 8 percent water. The solid matter is chiefly collagen matrix hardened by impregnation with calcium salts (Thomas 1985). Bones develop either from small cartilage models (anlage) in the eight-week-old embryo or from condensed embryonic tissue (mesenchyme) that forms a dense membrane (Arey 1966).

15

Facial bones and portions of calvaria, mandibles, and clavicles derive from the latter and are called "membranous," all other bones form in areas occupied by cartilage which they gradually replace and are therefore "cartilaginous."

Bone forms along the paths of invading blood vessels. Chondrocytes enlarge (hypertrophy) and proliferate (hyperplasia) about the blood vessels, becoming osteocytes as the cartilaginous matrix becomes mineralized. Enchondral ossification is a well-ordered sequential process of converting the cartilaginous model into bone. It is present under the perichondrium, a blood vessel rich layer of cells outside the model, and in active centers which develop within the bone. The fewer bones of membranous origin form along the many blood vessels developing within the membrane.

Both of the processes are referred to as *modeling* and any subsequent changes requiring resorption of preexistent bone followed by deposition of new bone is *remodeling*. Thus modeling is an early process, while remodeling occurs during normal growth and continues until death.

All bone is remodeled along a blood vessel advancing through cartilage or bone. Just ahead of this blood vessel is a cutting core of osteoclastic activity dissolving the bone. New bone is formed by the osteoblasts trailing beside the vessel. This is often referred to as a bone remodeling unit (Rockwood 1984).

Osteoblasts form the bone matrix osteocytes, serve to maintain it, and osteoclasts remove it. The ratio of osteoblasts to osteoclasts determines whether bone is deposited or resorbed. Since a single osteoclast destroys in 36 hours the same amount of bone that ten osteoblasts produce in ten days, it is obvious that if microscopic examination reveals an abundance of osteoclasts, bone resorption must be occurring (Snapper 1957).

Seven basic categories of disease may affect any sort of tissue or organ, and are best recalled by use of the mnemonics KITTENS:

K = Congenital/Genetic
I = Inflammatory/Infections/Idiopathic/Iatrogenic
T = Traumatic
T = Toxin
E = Endocrine/Metabolic
N = Neoplastic/Neuro-mechanical
S = Systemic

Congenital may refer to structural anomalies at birth, or genetic defects appearing later such as osteogenesis imperfecta. Inflammation of any sort, especially chronic infection, frequently causes remodeling. These changes may also be due to unknown causes or a poorly understood disease (idiopathic osteoarthritis) or caused by medical intervention (iatrogenic). Trauma is usually followed by reparative processes after initial resorption. Toxins may interfere with bone growth (lead poisoning) or become incorporated in the bone (e.g. lead, arsenic, tetracycline). Endocrine or metabolic effects vary markedly, from excessive growth of immature bones (gigantism) or the mature skeleton (acromegaly). Metabolic effects are most pronounced in nutritional deficiency states of vitamins (rickets, scurvy) and calcium deficiency (osteoporosis). Neoplasia may present as areas of resorption, with or without reactive deposition to produce a sclerotic margin. Neuro-mechanical defects such as tabes dorsalis from tertiary syphilis or the loss of sensation from diabetes mellitus destroy the control of smooth joint action producing traumatized points from the stressful gait. Systemic diseases like rheumatoid arthritis and lupus produce inflamed joints with recognizable patterns of destruction.

Osteoarthritis, however, is an example of a widespread inflammatory disease that is poorly understood and can be thought of as systemic (aging) or traumatic. The important point is that any classification system is not rigid but serves as a basis for organizing thought. Another factor to keep in mind as one holds the dry bone in hand is that extraordinary changes may not be due to a primary disease of bone, but the bone has been affected secondarily in the natural course of a disease as in sickle cell and hemophilia.

Description of the lesion is important in developing a list of differential diagnoses. Normal bone has a continual balance of resorption and formation and is subject to gradual remodeling of trabeculae to meet changes in stress loading. When stimulated, this balance is upset, but bone can react in only three ways (resorb, deposit or both). Recognizing the *predominant* process is the key to description. Once the pattern has been described by location, gross, and radiographic appearance, the number of disease possibilities can be narrowed.

For example, a single resorptive (lytic) lesion of a vertebral body could represent infection (e.g. tuberculosis), a primary cancer (plasmacytoma), or a benign lesion (Schmorl's node). If the lytic margin is surrounded by a reactive growth of bone the process is probably a chronic infection (e.g. tuberculosis). Aggressive lysis with no reaction favors malignant disease, while a smooth regular margin is probably both benign and quiescent.

Resorption of bone can also be caused by contact pressure erosion from tumors or along the path of dilated blood vessels. The latter occurs when an alternate pathway must be used after normal circulation is obstructed by congenital or acquired disease.

Deposition of bone is most frequently due to inflammation of the periosteum. This is most readily seen in the exuberant callus after fracture, but is also seen in more subtle injuries where muscular injuries next to the bone pull the periosteum away and blood collects underneath. The blood clot is organized into scar tissue which may be converted into bone. Chronic infections (osteomyelitis) of bone rarely heal spontaneously, producing years of inflammation resulting in very large deposits of bone, often with a central sinus which continually drained pus. Tumors, especially primary bone cancers produce wildly aberrant patterns of new bone admixed with areas of aggressive lysis. Hematologic disorders (thalassemia, sickle cell) are associated with a "hairy" change in the calvarial diploe as this typical X-ray finding represents expansion of the marrow space to compensate for the anemia. The trabeculae become vertically oriented, thus appearing like hairs standing on end.

The above are merely titillating examples of resorption, deposition or both, each depending upon interpretation of its pattern for diagnosis. This is why we study remodeling.

OSTEOARTHRITIS

A chapter on bone remodeling is not complete without a discussion of osteoarthritis (OA), also known as degenerative joint disease (DJD). Not only is OA chiefly recognized by remodeling changes in skeletal series, it is the most common manifestation of disease after dental caries. It is present in most persons older than 50 years, and in 90 percent of octogenarians.

As widespread as it may be, the causes of this disease are diverse and obscure. Osteoarthritis most frequently results in destruction of weight-bearing joints (vertebrae, hips, knees) although any joint may be affected. Two clinical patterns emerge, primary and secondary OA. Primary OA is idiopathic, of no known cause, but may be one of the many changes of aging. The spine is most often affected but the patient largely remains unaware of the process. Secondary OA is related to wear and tear. Traumatized joints (including repetitive micro trauma), congenital abnormalities, and other etiologies (remember KITTENS), are subject to the inflammation and repair process of any living, vascular tissues. Thus, the remodeling is limited to a specific degenerative process and hence the alternate term "degenerative joint disease."

A theory has been advanced to explain each clinical pattern. The biochemical theory seeks to explain primary OA as an aging phenomenon where the body gradually loses its ability to maintain joint cartilage. Stresses develop, damage the joint and the attendant inflammation in the synovium (synovitis) releases enzymes and inflammatory chemicals that attack the cartilage. Once the cartilage is damaged, the changes common to both clinical patterns follow.

A biomechanical theory emphasizes loading stresses directly injuring the cartilage cells, affecting their ability to maintain the matrix. The cartilage becomes soft, with fissures cracking the surface and flakes breaking off. The inflammation induced by this damage then chemically attacks the remaining cartilage and a cycle of accelerating damage develops. Changes in the underlying bone may accompany the process leading to microcysts beneath the cartilage which then collapses from loss of support.

Regardless of the mechanism, the final appearance is common to both. With the loss of chondrocytes and fragmentation of the softened cartilage, the subchondral bone becomes exposed. Exposed bone forms a small callus rich in blood vessels followed by an extensive remodeling producing thick polished (eburnated) bone resembling ivory. Other areas show osteoclastic activity with microcysts and even small fractures. The cartilage spreading out at the joint margins and the synovial lining turn into ossified outgrowths known as spurs (osteophytes). Chips of cartilage or spurs may roam at large in the joint and are called "joint mice."

The remodeled subchondral bone and osteophytes are obvious changes typical of OA. But hidden in the marrow cavity is another change. The trabecular pattern, which identifies the lines of stress, differs from the normal pattern and reflects changes in the stress loading of the bone. There is some debate as to whether this reflects cause or effect of the OA, whichever it is, radiographs soon reveal its existence.

Radiological and clinical features of OA in the living are: (1) narrowing of the joint space, (2) the osteophytes, (3) altered bone contour, (4) subchondral bony sclerosis and cysts, (5) periarticular calcification, and (6) soft tissue swelling (Dieppe et al. 1986). Most of these are helpful to the anthropologist. Questions anthropologists have raised in the past arose over the predictive value of OA findings in skeletal remains. Specifically, does OA reflect changes due to continuous stress or trauma (overuse) to a joint induced by characteristic activities? What are the limits of the normal "wear and tear" for old age? Can differences be found between the dominant and nondominant hands? These questions are particularly pertinent as physical anthropologists examine remains, devoid of soft tissue and lacking medical histories, trying to interpret what activities resulted in joint lesions. As the reader will see, no statement will be made that these questions can be adequately answered. We can only supply the reader with findings of several clinical studies focusing on the etiology and frequency of OA of contemporary groups.

A number of groups have reported comparing the incidence of OA in various occupations. Increased frequencies of OA were found in bus drivers and cotton pickers (Lawrence 1961, 1969), foundry workers (Mintz and Fraga 1973), print setters (from flicking letters across the tray with their thumbs, Dieppe et al. 1986), and coal miners (Kellgren and Lawrence 1958). However, pneumatic hammer operators (Burke and Fear 1977) and long distance runners (Puramen et al. 1975) that should expect to experience increased frequencies of OA, don't.

A study of 134 individuals aged 53–75 from Northern California addresses the hypotheses that increased frequencies of OA are associated with the dominant hand (93% were right-handed) or due to "wear and tear" (Lane et al. 1989). The subjects were studied using radiographs, rheumatologic evaluation, and questionnaire. Ninety-five percent of the individuals were classified as having occupations that were nonphysical. The subjects were separated into subgroups based on lifetime heavy hand use. The researchers found " . . . no significant differences or meaningful trends in any subgroup between dominant and nondominant hands." However, the authors concluded that since the subjects were in nonphysical occupations they may not have had enough chronic stress or trauma to accelerate joint degeneration in the dominant hand. This study as well as others showed that heavy use remains an open question. A broad range of wear may result in no adverse effects to the joints or the development of OA (Lane et al. 1989; Panush et al. 1986; Wright 1980). Thus, interpretations of skeletal remains deserves no less caution.

Readers wishing more detailed information on bone disease and remodeling should refer to Crelin (1981), Manchester (1983), Ortner and Putschar (1985), Steinbock (1976), or one of the many texts on embryology and pathophysiology.

Chapter V

DISEASES OF INDIVIDUAL BONES

SKULL AND MANDIBLE

Figure 1. Pathological conditions of the skull and teeth.

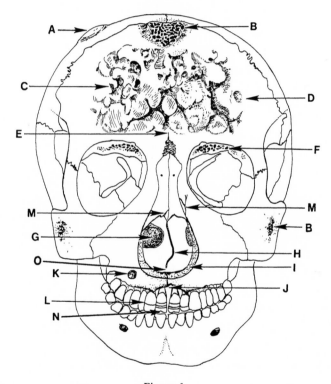

Figure 1.

A. Button osteoma (ivory osteoma)—small to large (ca. 1 cm), polished and roughly circular raised area of dense bone that resembles a small mound. These growths are classified as benign (harmless) tumors but, in fact, are not true tumors. Common occurrence in most populations. Ortner and Putschar 1985; Steinbock 1976.

B. Porotic hyperostosis (Angel 1966; El-Najjar et al. 1976; Lallo et al.

1977; Palkovich 1987), symmetrical osteoporosis (Zaino 1967), symmetric osteoporosis (Hrdlicka 1914)—small (0.5 mm) to large (2.0 mm) sieve-like holes involving the outer table and diploe accompanied by increased vault thickness. There is much debate over the etiology and proper classification of this condition. Some populations will show an extremely high frequency of porotic hyperostosis (e.g. coastal Peruvian), while other groups show little or none (e.g. contemporary American whites and blacks). Hrdlicka (1914), in examining 4800 crania, found porotic hyperostosis to be common among the prehistoric coastal peoples of Peru but absent in groups from the mountainous areas.

Regardless of the proposed etiology (e.g. iron deficiency or hereditary anemia [Miles 1975], nutrient losses associated with diarrheal disease [Walker 1986], vitamin or nutritional deficiencies [McKern and Stewart 1957]), toxic causes [Hrdlicka 1914], or infection, it is imperative that an accurate descriptive analysis be conducted. As the name implies, porotic hyperostosis should only be applied to those cranial bones that exhibit both porosis (holes) and thickened (increased) bone.

Many skulls will show tiny pits (porosity) of the parietals (most commonly), occipital, and frontal bone near bregma; however, no thickened bone will be present. Since porotic hyperostosis must by definition include the presence of thickened bone, the present authors have chosen to use the purely descriptive term ectocranial porosis for pitting of the outer vault giving it an "orange-peel" texture, not accompanied by thickened bone. Some researchers (Carlson et al. 1974; Lallo et al. 1977) believe that cribra orbitalia is an early form of porotic hyperostosis. Angel 1964; Dallman et al. 1980; Hill 1985; Lanzowsky 1968; Mensforth et al. 1978; Stuart-Macadam 1985, 1989).

C. Caries sicca (pronounced sick-uh)—diagnostic criterion of tertiary syphilis (lues; see Figure 7). Syphilitic skulls will exhibit mild to severe destruction of the outer table (radiographically, the inner table and diploe may also be involved). Caries sicca is comprised of raised, smooth nodules (nodes) surrounding singular or coalesced areas of bone loss (cavitation) that have undergone destruction and subsequent healing (sclerosis). Linear striations (stellate or radial scars) will be seen radiating from the nodes. The long bones should also be examined for evidence of remodeling. For a thorough description of the bony changes accompanying syphilis refer to Hackett (1976, 1981), Ortner and Putschar (1985), and Virchow (1896). For further historical information refer to Baker and Armelagos (1988).

D. Depressed healed fracture (injury)—frequent finding in American Indian crania. Such fractures may be found in any area of the skull but are most frequently seen in the frontal and posterior portions of the parietals. The typical appearance of a depressed fracture is a concave defect in the outer vault, with or without radiating fractures. The size of the wound is usually about the size of a dime or nickel and circular or ellipsoidal, although any size and shape may be encountered. Often it is difficult to distinguish a depressed wound from a healed lesion originating in the scalp. Walker 1989.

E. Remnant of the metopic suture (normal variant)—in a child the metopic suture (medio-frontal) separates the frontal bone down the midline. Normally the suture begins to obliterate at the end of the first year after birth with fusion being complete not later than the fourth to sixth year (Limson 1924). In some individuals (1% to 12% of adults. Krogman and Iscan 1986), the two frontals fail to unite resulting in "metopism" or a persistent metopic suture (nasion to bregma). Remnants of the metopic suture in the form of transverse irregular fissures above nasion are common findings while metopism is uncommon to common. Chopra 1957; Hess 1945; Latham and Burston 1966; Manzanares et al. 1988; Schultz 1918; Torgensen 1950.

F. Cribra orbitalia (Ursa orbitale. Moller-Christensen and Sandison 1963)—porosity and/or expansion of the superior orbital plates during childhood. This condition is usually bilateral and appears as small to large holes in the upper surfaces of the orbits. In children, the bone may actually be thickened while in adults only remnants of the holes (frequently only pits) remain. The frequency of this condition varies greatly by population and depends on a number of factors, many of which are under debate (e.g. iron-deficiency anemia, perhaps related to malnutrition, scurvy, chronic gastrointestinal bleeding, ancylostomiasis, and epidemic disease. Hirata 1988). Cribra orbitalia may or may not be accompanied by pitting and/or thickening of the outer table of the skull. Stuart-Macadam (1989) suggested that "The similarity between porotic hyperostosis of the orbit and vault with respect to macroscopic, microscopic, radiographic, and demographic features supports the idea of their relationship." Carlson et al. 1974; Hengen 1971; Steinbock 1976. (See Porotic hyperostosis.)

G. Pneumatized turbinate (nasal turbinate hypertrophy)—enlarged nasal conchae. The conchae appear swollen and rounded, possibly a bony response to allergy. Uncommon finding. Gregg and Gregg 1987.

H. Deviated nasal septum—common finding in most modern populations. The septum will deviate to either side of the midline of the nasal aperture and usually presents no clinical problem except when the deviation is so severe that it blocks the nasal passage. Uncommon in archaeological populations.

I. Eroded nasal spine and/or aperture—erosion of the nasal area is one symptom of syphilis and leprosy (facies leprosa), tuberculosis, and leishmaniasis among other conditions (Moller-Christensen 1961, 1974, 1978; Reichart 1976). Typically, erosion of the nasal spine is one of the early bony changes of the nasal area, followed by erosion of the inferolateral border of the aperture (rhinomaxillary change. Manchester 1989). Rare finding in most populations and uncommon to common in Europe (depending on the time period).

J. Alveolar resorption—common finding in lepromatous leprosy (leprosy that affects the skeleton). The roots of the teeth will become exposed due to bone loss of the alveoli. Care must be taken not to confuse erosion due to leprosy with that of simple periodontal disease (common finding), tooth loss from caries, or trauma. It is best to consult an oral pathologist when questions arise regarding the teeth or mouth. Ell 1988; Patterson and Job 1964; Manchester 1989; Moller-Christensen 1983.

K. Periapical abscess—inflammation at the apex of a tooth root with sinus formation (large to small pocket) and penetration of the maxilla or mandible. The margins of a periapical abscess (hole) will exhibit some periosteal reaction (pitting), and a pocket may be visible at the root apex. The margins of a healed abscess will be smooth, rounded, and of similar texture as the surrounding bone.

L. Linear enamel hypoplasia (hypoplastic lines, enamel dysplasia)— horizontal grooves resulting from a disturbance in the development of the enamel. Often, hypoplastic lines appear as deep grooves encircling the tooth crowns at the same level. Enamel hypoplasias can be caused by many factors including periapical inflammation or trauma to a deciduous tooth, fever, disease, nutritional deficiencies (especially A and D), endocrine dysfunction, and generalized infection during odontogenesis (Robinson and Miller 1983). A second form of hypoplasia, not shown, is represented by pits of various sizes in the enamel. Common finding in many populations (e.g. American Indian). Goodman and Armelagos 1988; Hutchinson and Larsen 1988; Pindborg 1970.

M. Fractured nasal bone and maxilla—common finding in many populations. Groups that practice(d) a lot of warfare or physical violence within the population may exhibit a high frequency of broken nasal bones and adjacent maxillae. Look for asymmetry of the nasal bones including depressions, small adhering bone fragments, and fractures with evidence of healing.

N. Eroded nasal spine—possibly reflects leprosy, tuberculosis, leishmaniasis, and syphilis among others. Rare finding. Manchester 1989; Moller-Christensen 1974.

O. Hutchinson's incisors, Hutchinsonian teeth, Hutchinson's teeth (Fiumara and Lessell 1970, 1983; Robbins 1968; Robinson and Miller 1983)—notched upper permanent central incisors (rarely the lateral incisors and canines may be notched) due to malformation of the middle gemma (Fiumara and Lessell 1970) at the time of morphodifferentiation (Robinson and Miller 1983). The defective enamel of the middle gemma leads to notch development after continued use. Hutchinson's incisors are diagnostic of late congenital syphilis, shorter than the lateral incisors, barrel-shaped or peg-shaped (tapered sides), and widely spaced in the tooth sockets. These incisors are usually thick both anteroposteriorly and mesiodistally (Fiumara and Lessell 1983). Also look for unusually shaped permanent molars (mulberry, Moon's, Fournier's), and a flattened nasal area (saddle-nose).

In a study of 100 individuals having congenital syphilis, mulberry molars were found in all of the patients who had their own teeth (some patients were edentulous or had their molars removed due to caries. Fiumara and Lessell 1983). Pavithran (1987) reported on a female with notching of only the left upper central incisor (unilateral). Care must be taken not to confuse Hutchinson's incisors with culturally modified teeth (Handler et al. 1982; Sweet et al. 1963) or the three small, normal undulations of the incisal edge known as mammelons. Hutchinson's incisors are rare in most skeletal samples. Steinbock 1976.

Figure 2. Pathological and normal conditions of the skull (lateral view).

A. Lacunae laterales (Grant 1972)—shallow endocranial depressions (raised "mound" ectocranially) located on either side of the sagittal suture. Although these depressions vary in size, most are large, smooth bordered, and serve to house large clusters of arachnoid granulations (Pacchionian granulations, Pacchionian bodies). Pacchionian pits, on the other hand, are small, sometimes clustered endocranial pits with sharply defined margins caused by erosion by small clusters of arachnoid granulations. Lacunae laterales are always located post-bregmatic (in the

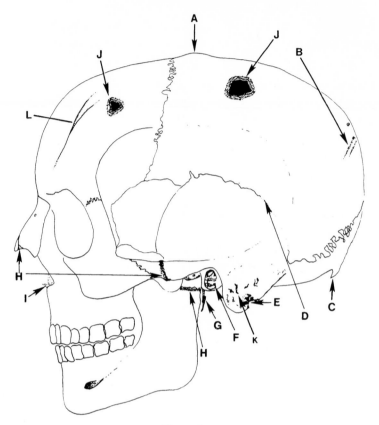

Figure 2.

anterior parietals only) while Pacchionian pits may be found in the parietals (possibly within the lacunae laterales) and frontal.

    There are a number of theories as to the etiology for the erosion of the inner vault of the skull due to arachnoid granulations. It is known, however, that these granulations primarily serve to filter and return cerebrospinal fluid. In some cases the lacunae laterales may become eroded and result in localized protrusion and perforation of the outer vault by the cauliflower-shaped and ossified arachnoid granulations. Both depressions (lacunae) and pits are extremely common in all populations and increase in number and depth with age. Perforations of the outer vault that are not due to post-mortem erosion are rare. (See Pacchionian pits.)

B. Normal vascular grooves—shallow grooves, usually symmetrically situated in both parietals, lateral and posterior to the parietal foramina.

C. Normal development of the nuchal crest—this area may either be flat (typically in females) or developed and projecting inferiorly in some males. Heavily developed nuchal crests may have a shelf-like ridge and an inferiorly-oriented "spike" (occipital torus) of bone in response to muscular activity and bone stimulation. Common finding.

D. Craniosynostosis (craniostenosis, premature suture closure)—term commonly used to refer to any suture that fuses at too early an age. While most authors use the terms craniosynostosis and craniostenosis interchangeably, others (Prokopec 1984) use the latter term only when the cranial capacity is diminished. When the sagittal suture prematurely fuses, the skull continues to grow and results in a long-headed individual (scaphocephaly or hyperdolicocephaly). Premature closure of the coronal results in a high (pointed) skull known as steeple skull, tower skull, or oxycephaly. Craniosynostosis does not refer to artificial cranial deformation from such practices as cradleboarding or head binding. Craniosynostosis is an uncommon finding in most skeletal populations and may be congenital, hereditary or the result of metabolic disturbances. Cohen 1986; David et al. 1982; Moore 1982; Prokopec et al. 1984; Stewart 1982.

E. Mastoiditis—inflammation and subsequent infection of the middle ear (otitis media) that may result in perforation and resorption of the mastoid process or other portions of the temporal bone. The middle ear is an air-filled space within the petrous portion of the temporal bone lying immediately behind the tympanic membrane (ear drum). Otitis media is very common in infants beyond the neonatal period (after 28 days of age) and shows a decline in incidence after the first year of life (Bluestone and Klein 1988). Skeletal involvement of the mastoid, however, is uncommon to rare in most skeletal samples. Care must be taken not to mistake the normal fissures in the outer surface of the mastoid for a pathological condition. Also be careful not to confuse the numerous air cells of the normal mastoid for disease. When mastoiditis is suspected, take a radiogram and look for sclerosis and pocket formation. Dugdale et al. 1982; Edwards 1989; Gregg and Gregg 1987; Paparella et al. 1980; Schultz 1979; Teele et al. 1984.

F. Auditory exostosis (torus)—benign bone tumor(s) of the ear canal. This tumor is easily visible as a rounded lump ranging in size from small (barely visible) to large, virtually filling the opening. Rarely the external auditory meatus may be absent at birth; no external canal will be seen in the temporal bone (the authors have encountered two such cases). Auditory exostoses are uncommon to common findings. Gregg and Gregg 1987; Kennedy 1986.

G. Healed fracture of the styloid process—a difficult fracture to detect because the normal styloid process may be irregular in shape, blunted or appear to have a portion missing. Fractures of the styloid process result from trauma to the neck or mandible (Haidar and Kalamchi 1980). Uncommon finding.
H. Fracture of the mandibular ramus (rare), zygomatic arch (uncommon), and nasal bones (common).
I. Erosion of the anterior nasal spine and nasal aperture (uncommon to rare).
J. Perforation due to disease of the cranium—diagnosing diseases based solely on examination of lytic lesions in the skull is extremely difficult. Look for new bone spicules encircling the lesion (both endo- and ectocranially), increased vascularity (tiny, smooth-bordered pits), and remodeling (healing or filling in) within the margin of the lesion. Some of the more common diseases that produce these lesions are metastatic carcinoma (cancer), tuberculosis, multiple myeloma, eosinophilic granuloma, and fungal infections (fungal infections and metastatic carcinoma can cause destruction of bone that is nearly identical to post-mortem damage). Seek the assistance of an experienced researcher before drawing any conclusions. Olmsted 1981.
K. Normal grooves and depressions in the mastoid process.
L. Accessory frontal sulci—one or more grooves above the orbits for transmission of branches of the supraorbital vessels and nerves (Grant 1948). Sometimes these shallow grooves trail into the supraorbital notch or foramen. Normal variant (nonmetric trait).

Figure 3. Pathological changes of the posterior skull and mandible.

A. Normal vessel grooves (common finding). (Large perforations or enlarged parietal foramina are the result of faulty ossification and are hereditary. Goldsmith 1922; Hoffman 1976; Miller and Keagy 1956.)
B. Biparietal thinning (Epstein 1953); senile atrophy (Wilson 1944)—symmetrical depressed areas in the posterior outer table of the parietals. The bone in these areas will be extremely thin, fragile, and translucent. The process of atrophy begins in the outer table, through the diploe and, in extreme cases, into and through the inner table (Wilson 1944). This rare condition is age related. The youngest affected individual known to the authors is 40 years of age. An interesting note is that the thinning usually avoids the parietal foramina leaving approximately 1 cm of undisturbed bone encircling these foramina (greatly enlarged foramina

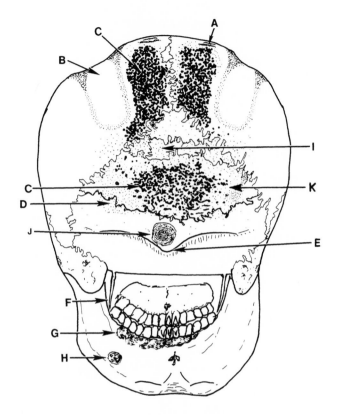

Figure 3.

[foramina parietalia permagna], may reflect a hereditary condition. O'Rahilly and Twohig 1952; Stallworthy 1932; Stibbe 1929; Symmers 1895). The presence of biparietal thinning will depend on whether or not some of the individuals in the population lived to old age. Lodge 1967.

C. Porotic hyperostosis (see Figure 1). Frequency varies depending on population. Angel 1964; Stuart-Macadam 1989.

D. Ectocranial porosis—tiny pits in the outer vault without increased vault thickness (see Figure 1). Common finding in most populations. Etiology unknown.

E. Normal development of the nuchal crest. Common finding in males, especially those who overuse the posterior neck muscles.

F. Elongated styloid process—elongation of the stylohyoid process due to ossification of the stylohyoid ligament; the normal length of the styloid process is 20–25 mm. Pain, impingement of the carotid artery, sore throat, and other symptoms associated with elongated styloids are referred to as Eagle's syndrome (Eagle 1948; Keur et al. 1986; Langlais et al. 1986; Monsour and Young 1986; Sivers and Johnson 1985). If this condition is suspected, seek the advice of an ear, nose, and throat specialist. DeChazal 1946; Douglas 1952; Dwight 1907a.

G. Torus mandibularis—rounded, usually symmetrical bony growths (tumors) along the superior lingual border of the mandible below the premolars. The tori may extend in the form of two bulging ridges that nearly touch one another behind the incisors. If the tori are quite large the tongue may have rested on them. Chronic gum disease (gingivitis), however, may also result in similar, albeit, less pronounced growths. Common finding (e.g. Eskimo) depending on the population.

H. Stafne's defect (static bone defect, posterior lingual depressions, lingual cortical mandibular defects, salivary gland defect, latent bone cyst)—(1) a circular or oval, smooth-walled concavity varying in size from 1–2 mm to approximately 1 cm in diameter and located in the lingual surface of the mandible inferior to the mylohyoid line or (2) a shallow, circular or oval and roughened defect less than 1 mm deep in the location noted above. On x-ray the defect has a sclerotic border. The location and radiographic appearance of these lesions are highly suggestive, although not pathognomonic (diagnostic), of Stafne's defects. Although the etiology of Stafne's defect is unknown, nearly all have proven to be benign defects, not tumors or cysts (Thawley et al. 1987), containing submandibular salivary gland tissue. The osseous defects result from pressure erosion of the mandible by the submandibular salivary gland or duct.

Preliminary examination (RWM) of over 5000 dry-bone mandibles from historic and prehistoric sites revealed 91 individuals with defects of which 81 were males (17 individuals from the same site in Alaska). Numerous researchers have confirmed the predominance of this trait in adult males (most individuals are in their forties or fifties when the defects are first detected). Stafne's defects are almost certainly developmental rather than congenital or traumatic in origin and may have a hereditary (sex-linked recessive) basis. The youngest known individual with a Stafne's defect was an 11-year-old boy from Sweden (Hansson 1980). Stafne's defects are usually unilateral. Because many of the defects are only 1–2 mm in length, radiographic detection in clinical cases may be difficult. Uncommon to rare finding in most skeletal samples. Correll et al. 1980; Finnegan and Marcsik 1980; Gorab et al. 1986; Harvey and Noble 1968–69; Pfeiffer 1985; Stafne 1942; Uemura et al. 1976.

I. Wormian bones (lambdoid ossicles, sutural bones)—small to large bones that may persist as separate ossicles or unite with the parietal and occipital bones.

J. Supra-inion depression (Stewart 1976)—a roughly circular depression in the occipital bone above the superior nuchal line. Although the etiology is unknown, there is a higher frequency of this trait in culturally deformed skulls (e.g. cranial binding) (Pers. Comm. Kathy Murray 1989). Uncommon to common finding.

K. Inca bone—a large accessory ossicle of the occipital bone. Although a true Inca bone is one that extends from asterion to asterion, most researchers identify any large accessory bone(s) in the lambdoidal region of the occipital bone as such.

Figure 4. Pathological conditions of the skull (inferior view).

A. Torus palatinus—a raised plateau of bone varying in size along the midline of the palate (not to be confused with the bony build-up along the anterior median palatine suture). The frequency and severity of this trait varies. Vidic 1966.

B. Peg-shaped third molar—smaller in circumference and of different shape than a normal molar. Congenital trait. Uncommon to rare finding.

C. Lesion in the basilar bone—rarely, osseous syphilis or tuberculosis may produce a lytic lesion (with a porous floor) in the basilar near the sphenoid bone (Ortner and Putschar 1985). The lesion may perforate the basilar and expose the sphenoid sinus. Care must be exercised not to confuse this lesion with the commonly occurring pharyngeal fossa (a smooth circular depression), a nonmetric trait.

D. Tympanic dehiscence (foramen of Huschké)—perforation of the tympanic plate of the temporal bone. The dehiscence appears as an irregular circular hole posterior to the temporal fossa (TMJ). This is a trait that is seen in all young children that only occasionally persists beyond age five years (Berry and Berry 1967). Uncommon to common finding.

E. "Tilt" head—twisted and asymmetrical occipital condyles and foramen magnum referring to a condition initially identified in Hawaiian skulls by Snow (1974). Premature closure of the cranial sutures (e.g. lambdoidal or coronal) may also result in an asymmetrical cranial base. Uncommon finding in most populations but frequently seen in Eskimos (experience of the authors).

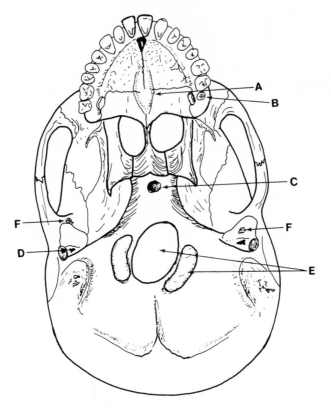

Figure 4.

F. Osteoarthritis of the temporomandibular joint (TMJ)—tiny to large pits and/or osteophytes on the articular surface and margin of the temporal fossa. Normal temporal fossae typically have undulating surfaces but will not have the porosity or bony build up. Since the TMJ is a paired joint that cannot function alone, OA, if present, is usually approximately symmetrical. Lesions are usually detectable earlier and are more severe in the temporal surface than in the mandibular condyle. If only one joint exhibits severe alteration look for signs of infection or trauma. OA of the TMJ is a common finding in most populations. Blackwood 1963; Markowitz and Gerry 1950; Ryan 1989.

Figure 5. Pathological and normal conditions of the skull (sagittal section).

A. Button osteoma in the frontal sinus. Common radiographic finding. Bushan et al. 1987; Dhooria and Mody 1986.

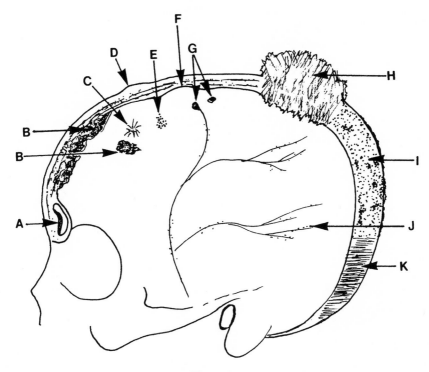

Figure 5.

B. Hyperostosis frontalis interna (Hyperostosis cranii. Moore 1955)—irregular, undulating bony growths located on the inner surface of the vault. Etiology unknown. Uncommon finding. Perou 1964.

C. "Radiating" lesions—small localized areas with radiating grooves directed away from a central depression. Usually there is no porosity or periostitis associated with these lesions. Etiology is unknown and may simply reflect increased vascularity. Uncommon finding.

D. Osteoma—small to large (ca. 1 cm in diameter), singular or multiple benign tumors. Ortner and Putschar 1985. Common finding.

E. Endocranial porosity—etiology unknown. May appear as tiny areas of microporosity that reflect inflammation in the outer table of the skull (Gregg and Gregg 1987). Uncommon finding.

F. Pacchionian/arachnoid/lacunae laterales erosion—erosion of the bone in the anterior portions of the parietals near bregma. These depressions are concave (when viewed endocranially), filled with arachnoid granulations, and exhibit gradually sloping margins. Probably a normal variant accompanying old age. The bone of the outer vault may be extremely thin, perforated (rarely), and bulged at these sites. Common finding in old individuals. (See Lacunae laterales.)

G. Pacchionian pits (arachnoid granulation pits; granular foveola. Waddington 1981)—pits resulting from erosion of the inner table of the vault due to enlargement and ossification of arachnoid granulations that serve to filter cerebrospinal fluid. In young individuals arachnoid granulations are villous and small. During old age the granulations enlarge, become cauliflower-shaped, and erode the cranial vault resulting in varying sized pits. Pacchionian pits appear as relatively small (2 mm) to large (5 mm) pits with sharply defined margins that are mostly confined to the parietals. The depth and frequency of these lesions increase with age and, possibly, disease. Common finding in all populations.

H. Hemangioma—Neoplastic growths (benign tumors) formed by a proliferation of blood vessels (Ortner and Putschar 1985). When viewed radiographically, the bone will appear to radiate from the center of the lesion in a "sun-burst" pattern. Rare finding. Spjut 1971; Steinbock 1976.

I. Paget's disease (Osteitis deformans. Paget 1877)—a chronic inflammatory condition that results in proliferation (thickening) and softening of bone that may affect any or all bones in the skeleton. Mirra (1987) hypothesized that a unique slow-virus infection of osteoclasts causes Paget's. In the early stages of this disease the lesions are typically lytic (resorptive), originate in one focus of bone, and slowly spread until the entire bone is affected (Mirra 1987). Late phases result in grossly enlarged, dense bones, especially noticeable in the skull and extremities. The disease seldom appears before the age of 40 years. Barry 1969; Ortner and Putschar 1985; Wells and Woodhouse 1975.

J. Pits along the meningeal grooves—although this condition may be normal in many individuals, it could represent a pathological state resulting in increased vascularity. Uncommon finding in most populations.

K. "Hair-on-end"—radiographs will show parallel rays of bone that resembles hair standing on end. The condition may reflect thalassemia major (Cooley's anemia), sickle cell disease, or other hereditary anemias. Hair-on-end is virtually diagnostic of anemia (Resnick and Niwayama 1981). Angel 1964; Moseley 1963.

Figure 6. Cross-section of the skull showing the typical shape of multiple myeloma and metastatic carcinoma (cancers).

A. Button osteoma—benign bony tumor in the frontal sinus. Common radiographic finding.

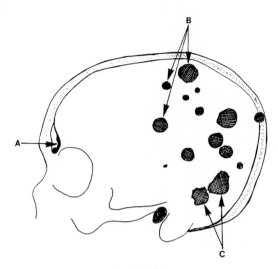

Figure 6.

B. Multiple myeloma (plasma cell myeloma, myelomatosis)—circular "punched-out" lesions of various sizes (Spjut et al. 1971). Uncommon to rare finding in archaeological samples.
C. Metastatic carcinoma—frequently, irregularly-shaped "moth-eaten" lesions of various sizes. Uncommon to rare finding in archaeological samples.

Caution must be exercised when making a differential diagnosis based solely on examination of the skull (carefully describe the appearance and distribution of the lesions throughout the skeleton). There are a number of other diseases that produce similar lesions including histiocytosis X (more correctly called Langerhans cell histiocytosis [Ladisch and Jaffe 1989]), multiple myeloma, fungal infections, and tuberculosis. The distribution of the lesions in the skeleton must be considered as well as their radiographic appearance, gender (prostate cancer in males and breast cancer in females; both forms, however, can result in blastic [dense] bones), and age. For example, if a child of less than two years of age exhibits these lesions, the acute phase of histiocytosis X (Letterer-Siwe) might be suspected while in adults, Hand-Schuller-Christian disease (the chronic stage of histiocytosis X) might be the proper diagnostic term (Lichenstein 1970). Bone involvement in histiocytosis X is 80 percent (Ladisch and Jaffe 1989). David et al. 1989; Jaffe 1972; Lichenstein 1953; Olmsted 1981; Ortner and Putschar 1985; Moseley 1963; Steinbock 1976.

Figure 7. *Tertiary syphilis of the frontal bone (upper drawing = ecto-cranium, lower drawing = endocranium).*

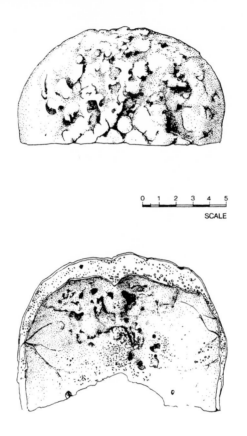

Figure 7.

Extensive destruction (cavitation) of the outer table and diploe, stellate scarring (radiating grooves), and nodules (mounds) in the outer vault caused by a variety of treponematoses. Note the smooth, raised, and rounded nodules (healed) as well as the depressed lesions that have eroded both the inner (rare) and outer tables. The late stage of syphilis (tertiary) can affect a number of bones but is most frequently seen in the tibiae (thickened periostitis and "sabre shin") and skull. Destruction of the skull may be evident in any area of the outer vault, maxilla, palate, malars, and nasal aperture (erosion of the nasal spine and border(s) of the aperture). Ortner and Putschar 1985.

Figure 8. Craniosynostosis (craniostenosis, premature closure of the cranial sutures).

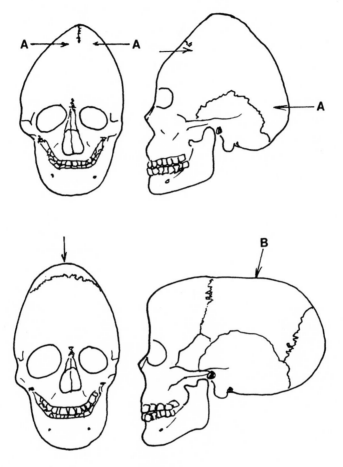

Figure 8.

Traditional terminology identifies and classifies the various forms of craniosynostosis, also known as cranial stenosis, craniostenosis (Tod and Yelland 1971), and premature suture closure, based on the resulting shape and deformity of the skull. Craniosynostosis does not, however, refer to cranial shape or deformation due to cultural practices such as cradle-boarding or head binding. David 1982; Moore 1982; Prokopec 1984.

A. Oxycephaly—also known as turricephaly, acrocephaly, hypsicephaly, tower skull and steeple skull. Although some researchers disagree, the term oxycephaly is usually used for skulls exhibiting premature closure of both the coronal and lambdoidal sutures. Perou (1964) states that oxycephaly is the most common form of craniosynostosis. Rare finding in most skeletal populations. (The authors have encountered this condition in only one skull.)

B. Scaphocephaly (hyperdolicocephaly)—condition resulting from premature closure of the sagittal suture. Scaphocephaly is one of the most common forms of craniosynostosis. (The authors have come across this form in seven crania, five of which were black). Some of the traits associated with this condition consist of a bulbous, projecting frontal bone and low-set eye orbits in relation to the frontal bone. Bilateral fusion of only the coronal suture is sometimes referred to as brachycephaly in the old literature. Unilateral fusion of the coronal is referred to as plagiocephaly.

Trigonocephaly (not shown)—premature closure of the metopic suture in utero resulting in a triangularly-shaped skull. Rare finding.

Premature lambdoidal closure (not shown)—fusion of only this suture is rare.

Plagiocephaly (not shown)—premature closure of one-half of the coronal or lambdoidal suture resulting in an oblique (skewed) skull (Perou 1964). Uncommon to rare finding (the authors have seen one case in an American Indian).

Figure 9. Macrocephaly.

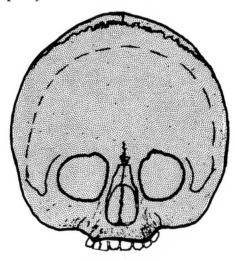

Figure 9.

This condition results in an unusually large and wide skull (the dashed line indicates the normal size of the skull). The cranial bones may be very thin and the fontanelles large. Macrocephaly may be congenital or due to a variety of other factors (e.g. hydrocephaly, "water on the brain"). Macrocephaly is a rare finding that is often difficult to discern from the proportionately large, yet normal skulls in children. Microcephaly (not shown) results in an abnormally small skull; also a rare finding. Zimmerman and Kelley 1982.

Figure 10. Depressed fractures (blunt force injuries) of the skull.

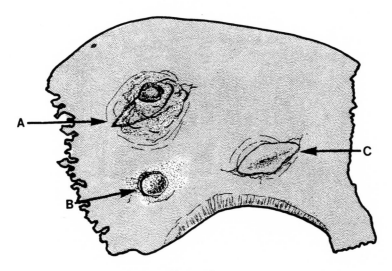

Figure 10.

Depressed fractures result from a variety of weapons and conditions. Although a large nonhealed (peri-mortem) fracture of the skull does indicate trauma to the head, it usually will not leave any evidence of the specific type of weapon used or the circumstances of the traumatic event(s). Further, healed depressed fractures may cause the investigator to suspect that the individual was struck by an assailant, but this may not be an accurate interpretation. For example, the person might have fallen from a rock ledge, struck his or her head, and died a week later. The fracture pattern could be identical if the person had been struck over the head with a blunt war club resulting in immediate death. Obviously, a projectile point embedded in bone leaves no doubt that some form of conflict occurred. What is important is that bony trauma must be carefully examined, described, photographed, and measured. The key phrase is "described and documented" with drawings and, preferably, photographs. Many times speculative interpretation is best not attempted.

Another important aspect of trauma is the pattern of the wound(s) in the population under study. For example, are all of the depressed fractures round? Where are the wounds located? How many sites of trauma show healing suggesting that the victim lived for some time after the event? Again, the overall population picture is important in reconstructing the events surrounding the trauma.

A. Healing depressed fracture of the right parietal—as you can see, it is difficult to determine whether the trauma at this site resulted from one blow or two. The central circular defect might represent one blow while the larger oblong defect another. Note the concentric fracture lines (A) surrounding the impact site(s). Close investigation might reveal some healing suggesting that the person lived after incurring the blow. Look for signs of infection and periostitis. It would also be a good point to measure (sliding calipers) the defects which might reveal that a similar shaped and sized weapon was used on a number of the individuals.

B. Circular depressed fracture—note the sharply defined margins around the central area of broken bone. Look for signs of dirt and uniform coloration within the wound suggesting that the breakage is old and not the result of post-mortem destruction. The edges around the depression may exhibit small adhering bone fragments that further suggest that the bone was broken while in a fresh state (look for "curled" or "curved" bone).

C. Depressed fracture with no healing (peri-mortem)—look closely for any indication of healing. If healing has occurred, there will be small patches of periosteal new bone growth and possibly porosity in the area of the defect, or along the fracture lines. If death occurred immediately or soon after sustaining trauma (up to about two weeks), the bone may show no new growth. As can be seen in the drawing, any number of objects or implements could have been responsible for the shape of the defect. The shape of the wound would depend, for example, on whether the person was hit with the pointed end of a club (small, circular depressed wound), or the blunt end (larger, oval or oblong depressed wound). Again, the pattern of the wounds in the population under study may help to clarify the type of weapon used. Current research by the authors on documented cases of skeletal trauma suggests that the term peri-mortem, when based solely on the amount of healing visible in bone, must be used with caution (osseous remodeling may not be grossly discernible until 10 days after sustaining trauma).

One of the authors (RWM) has found it very useful to shade in the shape of a skull wound using a piece of paper and pencil or artist's charcoal. If the depressed wound is in the frontal bone it would be "shaded" using the following method: make a drawing (approximate anatomical size) of the frontal view of the skull, note the approximate position of the wound, and correspond this with the drawing. Hold the drawing against the frontal bone and use the side of the lead to rub across the defect. The result is an unshaded (white) area above the depressed area (the lead won't come in contact with the concave areas). This method renders a silhouette of the size and, more importantly, the *shape* of the wound. You then can compare the depressed wounds in all of the skulls to see if a size and shape pattern is present which may reflect the weapon(s) used. Walker 1989.

Figure 11. Cranial alteration due to trephination (trepanation), rodent gnawing, and scalping.

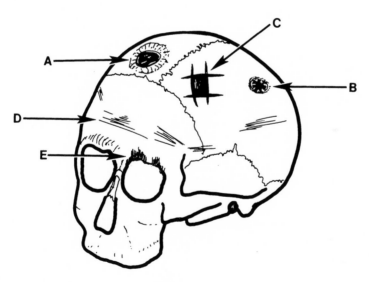

Figure 11.

A. Trephination is an ancient technique of removing a portion(s) of the cranium by scraping (A), drilling (B), or sawing (C). Although rare in North America, this practice is quite common in prehistoric skulls of

South America (in particular, ancient Andean populations), as well as many other groups throughout the world (John Verano Pers. Comm. 1989). Webb (1988) recently reported on two skulls from Australia that may represent the first such cases from that continent. Care must be exercised when trying to differentiate lytic lesions (e.g. infectious diseases, tumors), axe and sword wounds, and blunt trauma from trephinations in the cranium. Look closely for cut marks and abrasions around the defect which would indicate trephination. Healed trephinations will exhibit rounded margins and, possibly, sclerotic bone, tiny spicules or roughened (possibly porous) areas surrounding the defect where infection or healing has occurred.

Scalped skulls will exhibit patterned cuts that usually circumscribe the entire cranium (D). Typically the cuts extend across the middle of the frontal bone from temporal to temporal. Also look for tiny cuts on any raised area of the skull where the muscles are difficult to cut through (e.g. temporal lines, zygomatics, supramastoid crests, and mastoids). Rodent gnawing (E) may also be present in the skull as well as any of the long bones. Although it is sometimes difficult to distinguish rodent gnawing from defleshing marks (e.g. cultural practices before burial), the former are usually located above the eye orbits, nasal area, zygomatic arches, and mandible. Rodent gnawing appears as numerous short, *parallel* cuts localized to raised areas of the bone. A great deal of bone may be lost by rodent gnawing. Lisowski 1967; Shaaban 1984; Steinbock 1976.

Figure 12. Pathological changes (OA) of the temporomandibular joint (TMJ) (Markowitz and Gerry 1950).

A. Erosion and pitting (a) of the mandibular condyle—pitting and erosion of the condyle are common findings in all populations although the severity varies greatly from one group to another. Porosity appears as a localized area(s) of bone loss, usually with a sharply defined margin. The common sites for such destruction are noted in Figure 4. The dotted line denotes bone loss due to erosion or, simply, the wearing away of the joint surface. Alteration of the mandibular condyle may range from a very small area of pitting (porosis) to complete destruction of the articular surface. Diet, mechanical factors related to chewing, dental wear, caries (cavities), disc abnormalities, facial morphology, and abscesses all contribute to destruction of this joint. Richards, 1987, 1988; Richards and Brown 1981; Ryan 1989.

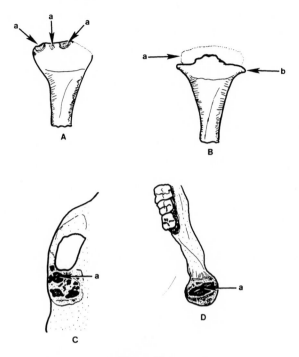

Figure 12.

B. Erosion (a) and osteophytes (b) of the mandibular condyle—when looking for erosion of the condyles, compare the shape and size of both heads for symmetry. There may be a little asymmetry in the condyles that is developmental or genetic and not representative of a pathological condition. Slight erosion is a common finding in most populations while severe forms appear to be age-related (elderly) or the result of trauma or chewing stress.

C. Osteoarthritis of the temporal fossa (temporomandibular joint and eminence)—small localized areas of porosity (a) in the temporal fossa are common in all populations. The first indication of porosity (pits) can usually be seen in the middle of the fossa or on the articular eminence (the raised area just anterior to the fossa; when the head of the mandible anteriorly dislocates, it rides forward and up on this eminence). Usually there is a well-defined rim with a central depressed area representing early OA. Moderate OA are those cases where large areas of bone are missing due to erosion (flattened) or porosity (a). Severe OA of the temporal fossa presents when most of the fossa has been eroded, with or without the formation of osteophytes.

Severe destruction and alteration of the mandibular head can result with little or no corresponding destruction of the temporal fossa. This latter point must be kept in mind when trying to match up an arthritic mandibular head with a questionable temporal fossa; the degree of destruction need not be similar. Richards 1988.

D. Osteoarthritis of the mandibular condyle—an example of severe destruction of the head in the form of pitting (a). "C" and "D" are from the same individual and are presented to show the degree of destruction of the two joint surfaces. Severe alteration of the condyle can result from trauma, infection or degenerative changes accompanying old age.

VERTEBRA

Figure 13. Osteoarthritis of the first cervical vertebra (atlas, superior view).

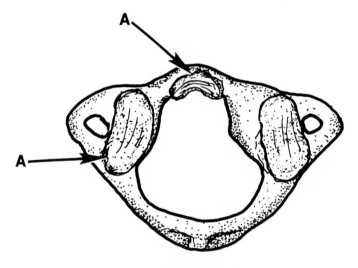

Figure 13.

Marginal osteophytes (A) that appear as irregular bony growths can be seen encircling the facets for the dens (odontoid process or peg) and articular facets for the occipital condyles at the base of the skull. Minimal development of osteophytes and porosity are common findings in the elderly. More severe OA will be evidenced by eburnation (polishing) and/or grooves in the articular surface of the dens facet. Lestini and Wiesel 1988; Sager 1969.

Figure 14. Osteoarthritis of the second cervical vertebra (axis, superior view).

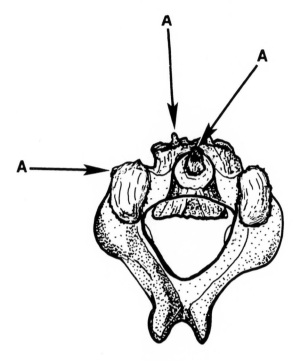

Figure 14.

The atlas may exhibit marginal osteophytes (A) and surface porosity on the articular facets or dens (odontoid process). In some instances an elderly individual will exhibit eburnation (polishing) of the anterior surface of the dens and ossification of the apical ligament (most superior portion of the dens). All of these changes are consistent with OA. Erosion of the dens, however, is one of the criteria associated with rheumatoid arthritis (RA). (In advanced cases of RA the entire dens may erode and fracture resulting in death of the individual.) The first cervical vertebra may also exhibit corresponding changes. OA is a common finding although unequivocable cases of RA, in archaeological specimens, are rare. Sager 1969.

Figure 15. Osteoarthritis of the cervical vertebrae (inferior view).

The most common finding of OA in the cervical vertebrae is marginal osteophytes (osteophytosis) of the anterior and posterior inferior margins of the body (A). Although the anterior border may be slightly

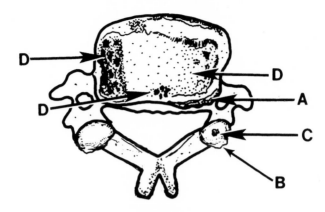

Figure 15.

irregular in normal vertebrae, osteophytes appear as bony extensions and spicules projecting inferiorly. Marginal osteophytes (B) and surface osteophytes (C) of the articular facets may also be present. In more severe cases the vertebral body may exhibit macroporosity and distortion of the end plates (D). Lestini and Wiesel 1988; Sager 1969.

Figure 16. Pathological conditions of the vertebrae (lumbar shown).

Before beginning a discussion of the spine, it is important to stress the difficulty in making a differential diagnosis and the complexity and confusing terminology associated with spinal disease. One condition of the spine may be referred to in the literature using many different terms. Further difficulty stems from the subtle and often similar bony changes in the early stages of many spinal disorders.

When examining a spine with fused vertebrae, one needs to differentiate ankylosing spondylitis, degenerative osteophytosis (spondylosis deformans), and diffuse idiopathic skeletal hyperostosis (DISH, Forestier's Disease). It is best to examine the vertebrae, take radiographs, describe the pattern of joint involvement, and seek the help of an orthopaedic radiologist. Hough and Sokoloff 1989.

A. Syndesmophyte—this type of bone growth, located at the margins of the vertebral bodies (centra) differs from the typical osteophyte associated with old age (commonly referred to as degenerative joint disease, osteophytosis, and osteoarthritis). Syndesmophytes are vertically-oriented growths of bone that form along the margins of vertebral bodies, within the annulus fibrosus, and accompany ankylosing spondylitis (AS), although other spondyloarthropathies (e.g. DISH) may present syndesmophytes.

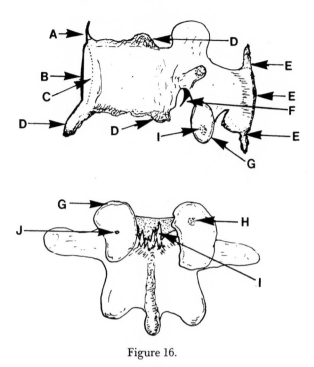

Figure 16.

Syndesmophytes consist mainly of calcified cartilage that is gradually replaced by regular bone tissue (Francois 1965). Uncommon to common finding.

Osteophytes, on the other hand, are roughly "C-shaped" (on cross section) bony projections at the vertebral margins that accompany old age and spinal trauma. If there is ossification of the paravertebral ligaments and spicules at many muscle and tendon insertion sites, the condition might reflect DISH. Osteophytes are a common finding. Dieppe et al. 1986; Ortner and Putschar 1985; Resnick and Niwayama 1988.

B. Squared anterior vertebral body—when viewed laterally, the normally concave body (C) is squared or flattened. This condition results from inflammation and erosion of the attachment of the intervertebral discs (annulus fibrosus) and vertebral body (Dieppe et al. 1986). Diagnostic of AS. Uncommon to common finding.

C. Original anterior body curvature.

D. Osteophyte—"beak-shaped" new bone that buttresses weakened vertebral bodies in the elderly (osteoporosis) and/or trauma to the spine. This is the form of bone growth that develops in degenerative joint disease. Common in all populations. Hilel 1962.

E. Ossification of the supraspinous ligament. Uncommon finding.

F. Neural canal spurs, para-articular processes (Hilel 1959), ossifying ligamenta flava (Grant 1972)—small bony projections located near the inferior articular facets of the thoracic vertebrae. Although this condition has been described as the result of ossified ligamenta flava, many researchers agree that these spurs represent a normal variant that is only pathological if they protrude into the neural canal far enough to impinge on the spinal cord (tethered cord) resulting in paralysis.

G. Marginal osteophytes on the articular facets. Common finding that increases in frequency with age.

H. Surface porosity (osteoarthritis) of the articular facets—common finding in all populations that shows an increase in frequency with age. In some instances a small, nearly circular defect (depression) might be seen in an articular facet that is not the result of OA. Such tiny depressions might be due to vessels entering the facet and are not pathological. The difference in the two defects (OA and vessel depression) are: the surface of OA porosities (within the defects) will be pitted and irregular, whereas vessel depressions will have smooth, compact floors and blunt margins.

I. Laminal spurs—spikes along the superior border of the "V"-shaped neural arch (lamina) that serve for attachment of the inferior ligamentum flavum (Hilel 1959). McKern and Stewart (1957) reported that laminal spurs were most often found in the thoracic vertebrae and, with rare exception, the cervicals. Normal variant of the spine. Common finding in all populations. Allbrook 1954; Davis 1955; Naffsiger et al. 1938; Shore 1931.

J. Normal nutrient pit for passage of the nutrient vessels into the facet.

Figure 17. Pathological conditions of the spine.

A. Squared anterior vertebral body—one criterion (diagnostic) of ankylosing spondylitis (AS). Radiographically (X-ray), the anterior portion of the centrum will be square and not show the normal concave curvature (Dieppe et al. 1986). This condition is commonly associated with ankylosis of the vertebral bodies and subsequent remodeling. Uncommon finding.

B. Hemivertebra (half-vertebra, lateral defect)—small, usually wedge-shaped and incompletely formed vertebra often with superior and inferior processes, and half a spinous process. A hemivertebra is frequently ankylosed to adjacent vertebrae and is a rare finding often associated with certain congenital deformities including Klippel-Feil syndrome and idiopathic scoliosis. Klippel and Feil 1912; Moe et al. 1978.

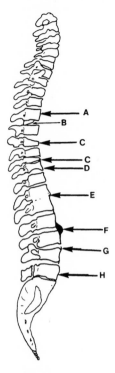

Figure 17.

C. Compression fracture (collapsed vertebra, wedged vertebra)—results from bone loss accompanying old age (osteoporosis), infection, or trauma such as from a fall from a rooftop. Compression fractures are responsible for the kyphosis (forward bending or hump back) in the elderly. The fracture and varying loss of centrum height most frequently results from collapse of the superior end plate, not the inferior plate. As the intervertebral disc protrudes into the end-plates, the bodies give way and collapse. Common finding in the elderly. Cummings et al. 1985; Raisz 1982.

D. Biconcave vertebra ("codfish" or "fish" vertebra)—due to protrusion of the firm but elastic intervertebral discs into weakened vertebral bodies. Severe osteoporosis is a main precursor to this condition. Common finding in the elderly, especially postmenopausal females (Albright et al. 1941). Twomey and Taylor 1988.

E. Ankylosis (anchylosis, fusion, coalition, bridging)—bony fusion of two or more contiguous vertebrae, usually along the anterior or lateral margins of the centra. The growths resulting in bony fusion can be distinguished as either osteophytes due to trauma or old age, or syndesmophytes (ankylosing spondylitis, Reiter's syndrome, DISH, and others).

F. Osteophyte (Dieppe et al. 1986) (osteophytosis, spondylitis deformans, spondylosis deformans)—horizontal, rounded bony growth commonly found in the elderly. Osteophytes result in small to large rounded protrusions of two or more adjacent vertebral bodies. These growths frequently appear as small, raised irregularities along the margins or midsections of the centra that gradually become larger until fusion of the bodies occurs. When two vertebrae become fused (ankylosed), the large osteophytes may be referred to as kissing or bridging osteophytes. Common finding. Hough and Sokoloff 1989.

G. Syndesmophyte (Dieppe et al. 1986; Resnick and Niwayama 1988) (one criterion of ankylosing spondylitis)—vertically-oriented bony growths that originate at the very margins (in the outer layers of the annulus fibrosus) of the vertebral bodies. These growths may result in large protrusions resembling osteophytes.

H. Spondylolysis (separate neural arch, Stewart 1956; bipartite lumbar, Grant 1972)—(pronounced spon"di-loli-sis. Thomas 1985) separation of the vertebral body from the posterior vertebral arch, usually at a junction known as the pars interarticularis or isthmus. Some researchers believe this condition to be congenital in origin while others state that stress plays a major role in causing the neural arch to separate (i.e., basically a stress fracture and nonunion). Activity involving the lower spine does seem to contribute to the presence of this trait. Most commonly the fifth lumbar is affected, but the fourth and third may also show this trait. Uncommon finding depending on the population being studied. Bradford 1978; Lester and Shapiro 1968; Nathan 1959; Ruge and Wiltse 1977; Willis 1924, 1931.

Figure 18. Pathological and normal conditions of the spine.

A. Bifid neural arch (spina bifida, split or cleft arch)—congenital malformation of unknown etiology resulting in incomplete closure of vertebral neural arches. When examining dry bone (no soft tissue or a patient history), it is extremely difficult, if not impossible, to determine if the condition was occulta (asymptomatic), cystica, or aperta (Dickel and Doran 1989 present an interesting case of a 15-year-old child dating to 7,500 years BP diagnosed as having aperta). Cystica implies that the skin was involved and the lesion became cystic while the term aperta refers to an open lesion (Strassberg 1982). The most common site of spina bifida is

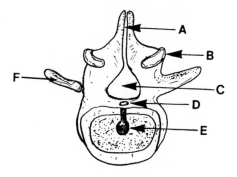

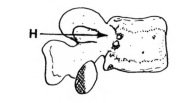

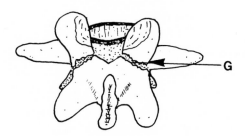

Figure 18.

S1 with a reported incidence of 9 percent in females and 13 percent in males (Cowell and Cowell 1976). Herniation of the spine can occur in even mild cases of spina bifida (James Vailas Pers. Comm. 1989). Ferembach 1963; Saluja 1988.

B. Marginal osteophytes of the superior or inferior articular facets. Look for raised ridges of bone that can be seen or felt with the finger or fingernail (marginal osteophytes are usually more irregular in shape than the normal facet margins). Common finding associated with old age and trauma.

C. Enlarged or diminished (stenosis, block vertebra) neural canal—in some instances the neural canal may be abnormally small, resulting in compression of the spinal cord (Resnick and Niwayama 1981). Stenotic lumbar neural canals (less than 10 mm anteroposteriorly) may be congenital or due to hypertrophy of the ligamentum flavum, vertebral body, and apophyseal joints (O'Duffy 1989). Rare finding. Eisenstein 1977.

D. Normal nutrient foramen.

E. Schmorl's depression (Schmorl's nodule, node, or cavity, disc herniation, cartilaginous node)—a circular, linear or combination of the two, depressed lesion, usually with a sclerotic floor in either of the centra end plates. In some cases only a small circular depression will be present in the center of the centrum—such lesions should also be scored as present. Schmorl's depressions result from expansion of the nucleus pulposus, the partially liquid central portion of the intervertebral disc. Expansion of the nucleus can take place in any direction but will bulge or herniate where the bone (frequently the end plate) or annulus fibrosus is weakest.

Schmorl's depressions are common findings in the elderly and result from degenerative disc disease. However, the presence of such nodules/depressions in adolescents is uncommon with only 2 percent of all herniated discs occurring in children and adolescents (Bunnell 1982). Schmorl's depressions in subadults result from trauma from such activities as a fall from height, heavy lifting, trauma during physical exercises, and similar activities.

Schmorl's depressions are common in all populations. When scoring its presence, the count can be based on either the number of vertebrae with the trait, how many total lesions are present in both the superior and inferior end plates, or the exact position of the lesions (e.g. between L-3 and L-4). For a good comparison of the frequency of herniated discs among populations see Thieme (1950). Bulos 1973; Gibson et al. 1987; Campillo 1989; Giroux and Leclerq 1982; Lindblom 1951; Saluja et al. 1986; Schmorl and Junghanns 1932.

F. Separate transverse process—a condition resulting in a divided transverse process and a smooth-surfaced articular facet located just inferior to the superior articular facet, at the base of mammillary tubercle (a lumbar rib attaches to the centrum and a separate transverse process to the neural arch). The authors have encountered one unilateral case in a first lumbar vertebra. Rarely a transverse process, usually a lumbar vertebra, may fracture and separate (Hoppenfeld 1989). Rare finding.

G. Spondylolysis (separate neural arch)—separation of the neural arch, typically at the pars interarticularis (also known as the laminae or isthmus), although separation may rarely occur through the pedicle. The usual site of spondylolysis is the fifth, fourth and, rarely, the third lumbar vertebra. (Added attention is given to the topic of spondylolysis due to its interest and controversial nature by many anthropologists.)

Although the etiology of spondylolysis remains unresolved, factors including genetic, congenital, and trauma (Blackburne and Velikas 1977; McKee et al. 1971) to the back must be considered. Thieme (1950) hypothesized that spondylolysis results during the developmental stage to final fusion. The lack of normal repair callus at the lesion, even in unilateral instances where movement of the "separating" arch is held at a minimum, suggests against postfusion breaks. Garth and Van Patten (1989) reported a case of bilateral fracture of the lumbar laminae not at, but near, the pars interarticularis that showed no bony union after two months.

The frequency of spondylolysis varies by population reaching as high as 40 percent in Eskimos (Stewart 1953). Libson et al. (1982) reported that of 1598 patients with lumbar spondylolysis, only two individuals had lesions in the upper three lumbar vertebrae. However, spondylolysis in the lower two lumbar vertebrae is more common. Rowe et al. (1987) found the defect in the lower lumbar in 2 percent to 10 percent of active young individuals in the United States. The male to female ratio is 2:1 from age six to adulthood (Fredrickson 1984), and the condition has never been reported as present in newborns. The youngest individual on record with spondylolysis is a 3.5-month-old infant (Seitsalo et al. 1988). Care should be exercised not to confuse spondylolysis with post-mortem breakage which is common at this site. Spondylolysis is a common finding and is strongly associated with spina bifida. Bradford 1978; Bridges 1989; Eisenstein 1978; Floman et al. 1987; Fredrickson et al. 1984; Griffiths 1981; Harris and Wiley 1963; Hitchcock 1940; Hoppenfeld 1989; Lamy et al. 1975; Laurent and Einola 1961; Letts et al. 1986; Libson et al. 1982; Lowe et al. 1987; Miles 1975; Nathan 1959; Newman 1963; Ravichandran 1981; Rosenberg et al. 1981; Stewart 1953, 1956; Thieme 1950; Troup 1976; Wiltse 1975.

If the vertebral body slides forward (which may be detected by examination of bridging osteophytes), the condition is known as spondylolisthesis (Congdon 1931; Merbs and Euler 1985; Pedersen and Hagen 1988; Ruge and Wiltse 1977; Thieme 1950; Wiltse 1962; Wiltse et al. 1975).

H. Lumbar rib—lumbar ribs are approximately 2–3 cm in length and have a single oval-shaped depression for articulation with the vertebral body. Look for a small single or double articular facet on the centrum where the rib was attached. Lumbar ribs may be bilateral and can occur in thoracic and lumbar vertebrae (not to be mistaken for a cervical rib). Three cases have been encountered by the authors, all in contemporary forensic cases that involve the entire transverse process (11th and 12th thoracics bilaterally; 1st lumbar in two individuals, courtesy of Lee Meadows). Rare finding in archaeological samples.

Figure 19. Bony changes associated with ankylosing spondylitis (AS), and degenerative joint disease (spondylosis) of the spine. Hough and Sokoloff 1989; Meisel and Bullough 1984.

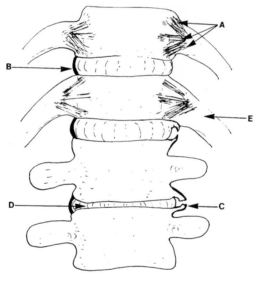

Figure 19.

AS is a chronic inflammatory disorder of unknown etiology. The prevalence of AS varies by population from absent in black Africans and Australian aborigines to 4.2 percent in adult male Haida Indians (Masi and Medsger 1981). Contemporary populations show an incidence of about 1 per 1000 individuals with a male to female ratio varying from 4:1 to 10:1 (Resnick and Niwayama 1981). The typical age of onset is 15 to 35 years but children may also be affected (juvenile-onset AS). Bony changes typically begin in the sacroiliac region (sacroilitis and subsequent ankylosis) and extend up the spine. As the spine becomes more involved it takes on an undulating contour ("bamboo spine") due to the development of extensive syndesmophytes between the vertebral bodies (Resnick and Niwayama 1981). AS often results in ossification at tendinous and ligamentous attachment sites (enthesophytes) to bone and in some cases may be difficult to distinguish from DISH. For further information refer to Dieppe (1986) and Resnick and Niwayama (1988).

A. Ossified tri-radiate ligaments resulting in fusion of the ribs to the vertebrae are typical of ankylosing spondylitis. Uncommon to rare finding.

B. Syndesmophyte (endesmophytes)—vertically-oriented bone growth typical of AS that forms from within the *margins* (annulus fibrosus) of the vertebral bodies (Francois 1965). Common finding, depending on the population under study. Resnick and Niwayama 1988.

C. Osteophyte—horizontally-oriented, rounded bone growth associated with degenerative spinal disease (osteophytosis), increasing age, and trauma. Common finding in all populations.

D. Intervertebral disc—gelatinous, fibrous and elastic discs that separate the vertebrae and serve as the shock absorbers of the spine. Epstein 1976; Gibson et al. 1987; Giroux and Leclerq 1982; Schmorl and Junghanns 1932, 1971.

E. Rib.

Figure 20. Diffuse idiopathic skeletal hyperostosis (DISH, ankylosing hyperostosis, Forestier's Disease) of the spine.

Although the etiology is unknown, DISH is a common disease in middle-aged and elderly individuals with males affected 2:1 over females (Utsinger 1984). Although DISH frequently affects the spine, other peripheral skeletal sites may be involved and exhibit "whiskers" (periostitis. Utsinger 1985) or irregular ossifications known as enthesophytes (spikes or projections) at tendon and ligament attachment sites (e.g. iliac crest, ischium, greater and lesser trochanters, trochanteric fossa, patella, calcaneus, ulna, linea aspera).

Criteria (Resnick and Niwayama 1976) for distinguishing DISH from AS include: (1) flowing calcification and ossification along the anterolateral surfaces of at least four contiguous vertebrae, (2) relative preservation of intervertebral disc space in the involved segments, and (3) the absence of apophyseal joint (facets) bony ankylosis and sacroiliac erosion, sclerosis or osseous fusion. Utsinger (1985) revised the diagnostic criteria of Resnick and Niwayama as follows: continuous ossification along the anterolateral surfaces of at least two contiguous vertebral bodies, primarily

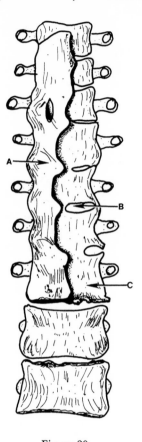

Figure 20.

in the thoracolumbar spine (broad, bumpy, buttress-like band of bone) and symmetrical enthesopathy of the posterior heel, superior patella or olecranon.

Rarely ankylosing spondylitis and DISH coexist. An extensive radiographic survey of 8993 persons over the age of 40 by Julkunen et al. (1973) revealed a standardized prevalence of 3.8 percent in men and 2.6 percent in women. Interestingly, the Pima Indians of Arizona showed an incidence of 34 percent in males and 6.6 percent in females (Henrard and Bennett 1973). AS is inherited as a Mendelian dominant trait with 70 percent penetrance in males and 10 percent in females (Sharon et al. 1985). Uncommon to common finding.

A. Flowing osteophytosis—thickened, flowing ossification along the right anterolateral aspect of the thoracic vertebrae (left side of the spine is usually spared due to the presence of the aorta). Note that the "ribbon" of ossification is clearly visible, extends across at least four contiguous

vertebrae, and appears as if a thick, undulating, and smooth surfaced band of bone has been applied to the right side of the spine. Some of the thickened ribbon of bone is the ossified anterior longitudinal ligament.
B. Normal disc space is preserved.
C. Twelfth thoracic vertebra.

Figure 21. Spondylosis deformans (spondylitis deformans, degenerative hypertrophic spondylitis, osteoarthritis, degenerative spondylosis) of the spine.

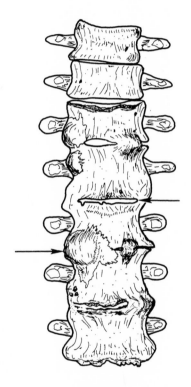

Figure 21.

Spondylosis deformans (SD) is the most common degenerative spine disease affecting males much more frequently than females, is occupation related, and is found in nearly all individuals over the age of sixty. SD is characterized by small to large bridging osteophytes (arrows) that bulge at the level of the intervertebral disc and serve to reinforce the centra. The proper use of the term osteoarthritis refers only to degeneration and involvement of the apophyseal joints (articular facets) whereas osteophytosis refers to osteophyte formation along the vertebral bodies and degenerative disc disease (Hough and Sokoloff 1989). The following description given by Norman (1984) best illustrates the process of osteophyte formation commonly associated with old age:

> The early changes of spondylosis deformans affect the anterolateral margin of the vertebrae where the annulus fibrosus inserts. Tearing of the fibers weakens the annulus. The restraints on the nucleus pulposus are lost, and the disk protrudes forward. Further stress will lift the anterior longitudinal ligament from the vertebral body, and a buttress of periosteal new bone fills in the area of separation. The osteophytes enlarge in a horizontal direction and curve to bridge the intervertebral disk space . . .

Many researchers attribute osteophytosis to many years of wear-and-tear that necessarily accompanies old age (Keim 1973). However, trauma, heavy physical stresses to the spine, and obesity may also result in osteophyte formation. Dieppe et al. 1988; Hough and Sokoloff 1989; Jaffe 1975; Ortner and Putschar 1985; Steinbock 1976; Trueta 1968.

Figure 22. Ankylosing spondylitis ("bamboo" spine, "poker" spine, Marie-Strumpell disease).

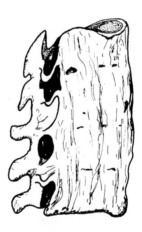

Figure 22.

This example shows fusion and the typical flowing appearance of the cervical vertebrae (syndesmophytosis) reflecting AS (see Ankylosing spondylitis). Spinal diseases are extremely difficult to differentiate and usually require the assistance of a specialist. Calin 1985; Dieppe et al.

1988; Ortner and Putschar 1985; Resnick and Niwayama 1988; Steinbock 1976.

Figure 23. Aortic aneurysm of the vertebrae.

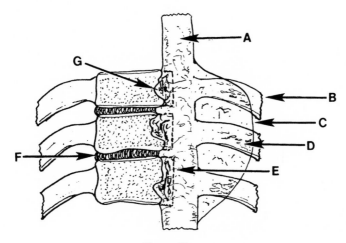

Figure 23.

Aortic aneurysms result in large, scooped-out lesions of the vertebral bodies caused by positive charges of the exterior surface of an expanded aorta. Typically, males over the age of 50 are affected (Keim 1973). The lesions are restricted to the left anterolateral surface of the spine. Distinguishing aneurysmal from tubercular erosion in an isolated vertebra is difficult. Rare finding in most populations (infrequent finding in cadaver and necropsy collections).

A. Aorta.

B. Ribs.

C. Aneurysm as it enlarges and comes in contact with the pleural surfaces of the ribs.

D. Small well-circumscribed excrescences of new bone, possibly responding to irritation by the greatly enlarged aorta.

E. Dashed line indicates the original width of the vertebral bodies.

F. Intervertebral disc.

G. Eroded portions of the vertebral bodies due to expansion of the aorta.

Figure 24. Three common forms of spinal deformity.

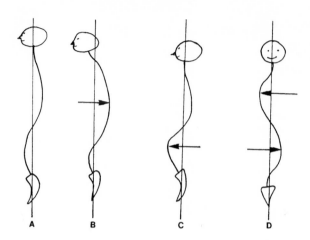

Figure 24.

Note that the drawing depicting a normally curved spine (A), kyphosis (B), and lordosis (C) are viewed from the side, while that of scoliosis (D) is seen from the front. These conditions can only be detected by articulating the vertebrae in the approximate anatomical position (align the articular facets). Remember that each intervertebral disc accounts for approximately 0.5 to 1.0 cm space (in young to middle-aged individuals) between adjacent vertebrae. Kyphoscoliosis (forward and to the side bending) is a common finding in the elderly. The dowager's hump of the elderly, usually most marked in females, is a common form of kyphosis. With increasing age the vertebrae lose bone mass (osteoporosis), collapse, remodel, and become anteriorly wedge shaped. Physiological scoliosis in which the vertebral bodies are greatly distorted and asymmetrical is a rare finding. A number of etiologies may result in scoliosis including tuberculosis, nonspecific osteomyelitis, acute trauma, osteoporosis, or osteoid osteoma (Haibach et al. 1986; Savini et al. 1988; Swank and Barnes 1987). Davis 1955.

Figure 25. Biconcave vertebra ("codfish," "fish" vertebra).

Biconcave vertebrae are the result of osteoporosis and expansion of the intervertebral discs into weakened end-plates of the bodies (not to be confused with "butterfly" vertebra, a congenital malformation of the vertebral bodies). Although acute trauma may cause fracturing and concave vertebral bodies (arrows) in a young individual, biconcave vertebrae are typical of postmenopausal women who lose bone mass (osteo-

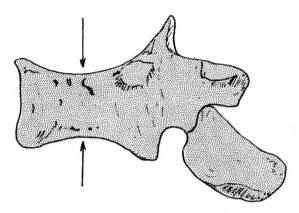

Figure 25.

porosis) much quicker than do males. Uncommon finding depending on the population under study (e.g. common finding in contemporary populations that live to be 60 or 70+ years). Dieppe et al. 1986.

Figure 26. Scoliosis of the spine (lumbar used as example).

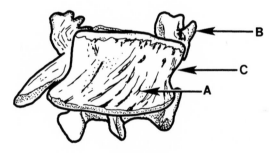

Figure 26.

Scoliosis is defined as "one or more lateral-rotatory curvatures of the spine" (Keim 1972). Idiopathic, or genetic scoliosis accounts for approximately 70 percent of all scoliosis and is divided into infantile, juvenile, and adolescent. Scoliosis is a common finding in the elderly, especially women (postmenopausal or senile osteoporosis) above 70 years of age (this form, however, is due to osteoporosis and collapse of the vertebral bodies and is not congenital in origin). Moderate to severe scoliosis in young individuals is a rare finding in most skeletal samples. Raisz 1982; Trueta 1968.

In examining a spine for physiological scoliosis look for "twisted" (asymmetrical) vertebral bodies (A), irregularly shaped and positioned articular facets (B), and missing transverse processes (C). The upper half of the centra may appear to be shifted in one direction while the lower half is shifted in the opposite direction. When the vertebrae are correctly articulated the spine will spiral, sometimes to an extreme degree. Severe scoliosis is associated with a number of pathological conditions including cerebral palsy, Marfan's syndrome, neurofibromatosis, and Klippel-Feil syndrome (Klippel and Feil 1912; Prusick et al. 1985). Dickson 1985; Farkas 1941; Moe et al. 1978.

Figure 27. "Butterfly" vertebra (sagittal cleft vertebra; oblique and superior views).

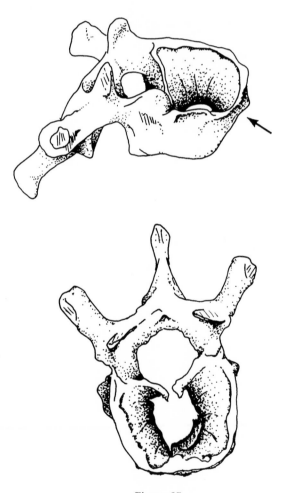

Figure 27.

Failure of fusion of the lateral halves of the vertebral body (Edelson et al. 1987; Epstein 1976; Moe et al. 1978; Pfeiffer et al. 1985). This rare condition is classified as a congenital defect of formation (developmental anomaly in the early embryonic period. Muller, O'Rahilly, and Benson 1986) resulting in a variety of malformed centra. The body of the vertebra may be cleft down the midline resulting in two hemivertebrae or thin strands of bone may transverse the cleft (Epstein 1976). Other bony changes may include widely spaced pedicles, malformed neural arches, and marked disarray of the posterior elements. The intervertebral disc may bulge into the defect. Butterfly vertebrae result in kyphosis (forward bending) due to the diminished height of the anteriorly-wedged vertebral body. Note the narrow, wedge-shaped anterior body (upper drawing, arrow) and cleft posterior vertebral body. (Subadult thoracic vertebra from Peru used as an example, courtesy of John Verano.)

Figure 28. Tuberculosis of the spine (superior view of a lumbar vertebra).

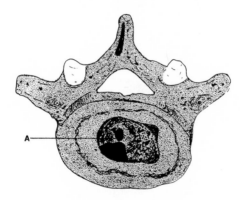

Figure 28.

A. Tuberculosis of the spine (Pott's disease, vertebral tuberculous spondylitis) showing the typical cavitation of the vertebral body (A). Destruction of the vertebrae usually first begins in the anterior, inferior portion of the vertebral body near an intervertebral disc, ultimately destroying the entire body and resulting in its collapse. Localized destruction (cavitation and abscessing) of the vertebral bodies (see Actinomycosis for comparison) is an inflammatory response to invading tubercle bacilli resulting in the formation of tubercles that stimulate erosion of the trabeculae and cortical bone. Note the minimal new bone formation although a great deal of bone has been resorbed. Adjacent vertebrae may show destruction where the inflammatory stimulus spreads beneath the anterior longitudinal ligament (sometimes the disease will "skip" verte-brae and continue above or below the initial vertebral abscess). Other frequent sites of tubercular involvement are the inner surfaces of the ilium (psoas abscess through the spread of the disease along the psoas muscle), skull, joints (usually only one joint is affected), and ribs. Uncom-mon finding depending on the population under study. Kelley and El-Najjar 1980; Morse 1961, 1969; Ortner and Putschar 1985.

Figure 29. Spinal tuberculosis (Pott's disease) with abscess formation of the vertebral bodies (lateral view).

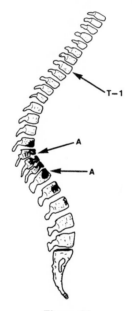

Figure 29.

T-1 denotes the first thoracic vertebra and "A" the diseased vertebral bodies. The darkened portions of the centra represent bony destruction that would soon result in collapse and fusion of the bodies and a forward bending/angulation (gibbus) of the spine. Tuberculosis of the spine has sometimes been confused with actinomycotic and mycotic infections which present a different pattern of spinal involvement (Simpson and McIntosh 1927). Examination of the spine reveals that the original location of the disease was "A" with subsequent abscess formation extending both above and below this central focus. Typical pattern of involvement in long-standing vertebral tuberculous spondylitis. Nonspecific osteomyelitis of the spine must also be considered. Morse 1961, 1969; Ortner and Putschar 1985.

Figure 30. Osteomyelitis of the spine.

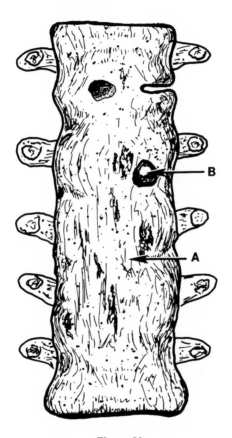

Figure 30.

Osteomyelitis of the spine is an uncommon finding. Without close scrutiny it may be difficult to distinguish between nonspecific osteomyelitis and tuberculous osteomyelitis (Steindler 1952). There are, however, criteria that help to make the distinction: osteomyelitis typically results in considerable bony reactive proliferation (A) while tuberculosis exhibits very little reactive bone formation (Carney and Wilson 1975). Both diseases may produce draining sinuses (cloacae) (B), but tuberculosis commonly results in destruction of the vertebral bodies, collapse and union of adjacent vertebrae (gibbus), and anterior deformity/bending and hump back (kyphosis). Gibbus formation is rare in osteomyelitis. Osteomyelitis seldom affects the transverse arches while tuberculosis does. Both are uncommon to rare findings. MacAusland and Mayo 1965.

RIB

Figure 31. Pathological conditions of the ribs.

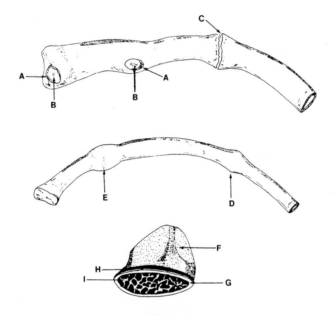

Figure 31.

A. Marginal osteophytes of the articular facets (OA). Raised, irregular bony growths where the rib attaches to the transverse process or facet on the centrum. Common finding accompanying middle and old age.
B. Surface porosity. Common finding.
C. Pseudarthrosis (pseudoarthrosis, malunion, false joint)—an irregular joint that results when movement continues between two fracture edges (complete fracture). For example, if the shaft of a radius is broken in two and the opposing ends are allowed too much movement during the healing stage, the ends may fail to unite. For union (callus formation) to occur, the fracture site must be either positioned in a manner so as to restrict movement of the broken ends (e.g. immobilized in a cast), or

motion of the broken ends must be minimized (e.g. primitive splints). In some cases when the broken ends fail to properly unite, cartilage develops instead of bone (callus), and a pseudarthrosis results. Movement of this false joint may properly function for many years after its formation. Uncommon to rare finding.

D. Pseudofracture—in some individuals there will be enlarged areas of bone near the sternal end of the ribs. Although these areas resemble healed fractures, look for similar "fractures" in both the left and right ribs. If all of the "fractures" are located in the same general area in many ribs and are of the same shape and size, they probably represent normal variation for attachment of muscles (other possibilities include myositis ossificans and benign osteophytes). Common finding.

E. Healed fracture—may be a complete or incomplete fracture. Look for swelling at the site and, possibly, sharp spicules or projections of bone. Radiographs may also help to distinguish a fracture from normal variation. Common to uncommon finding.

F/H. Active or healed periostitis—enlarged and thickened periosteal bone confined to the visceral surfaces of the ribs. Small, roughly ovoid and well-defined areas of reactive bone may be indicative of tuberculosis (Marc Kelley Pers. Comm. 1989). The etiology of ribs that are greatly thickened is unknown although tuberculosis and other chronic pulmonary diseases are suspected. Carefully examine the entire skeleton (especially the spine) for other indications of infection. Uncommon to rare finding in archaeological samples but common in early twentieth-century specimens (Charlotte Roberts Pers. Comm. 1989).

G. Original cortex (cross section of a rib).

I. Trabeculae (spongiosa)—the internal architecture of a rib. Normal finding.

Figure 32. Probable tuberculosis of the ribs (dorsal or visceral surface).

A. Articular facet for articulation with the transverse process.

B. New bone (thickened) possibly resulting from spread of the infection to the ribs or mechanical irritation due to pleural expansion (Marc Kelley Pers. Comm. 1989). The surface of the affected ribs may be irregular, undulating, and pitted. Independent research by Marc Kelley, Keith Manchester, and Charlotte Roberts is currently being conducted to determine the etiology of these lesions since their exact cause is unknown. Uncommon to rare finding depending on the temporal and geographical origin of the sample being studied.

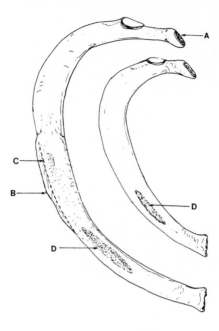

Figure 32.

C. Dashed line indicates the original cortex before new bone has thickened the rib; tiny pits may also be present.

D. Small, well-circumscribed areas of roughened bone (periostitis) indicative of early rib involvement. The patches appear as oblong, slightly raised areas of new bone following a pattern of increasing and decreasing size as you go up or down the articulated rib cage (Marc Kelley Pers. Comm. 1988). The ribs of infants and children should be closely examined for such lesions. Frequently the vertebral ends of the ribs will be most affected. Kelley and Micozzi 1984.

Figure 33. Bifurcated (bifid. Grant 1948) right rib (inferior view).

A single rib that bifurcates (divides) into two near the sternum (Popowsky 1918). Note the normal single rib shaft and articular facets for articulation with the spine. This condition most commonly affects the third or fourth rib as it divides (bifurcates) near the sternum (Grant 1972). Uncommon finding. Bloomberg 1926.

A. Single rib.

B. Bifurcated sternal ends.

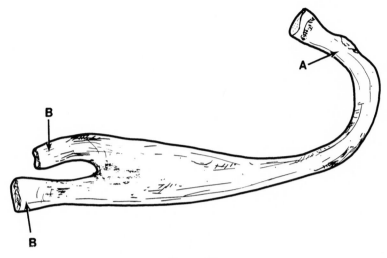

Figure 33.

Figure 34. Bicipital rib (right first and second as example).

Figure 34.

Two ribs that fuse to form one. Bicipital and bifurcated ribs differ in that the former are true fusions of what should have been two independent ribs, usually those of the first and second thoracic vertebrae. Fusion of the two rib shafts (bodies) is complete at the sternal end. Uncommon to common finding. Not to be confused with congenital fusion of many ribs associated with certain syndromes (e.g. Klippel-Feil).

A. Independent vertebral portion of one rib.
B. Independent vertebral portion of a second rib.
C. Common shaft formed by fusion of two rib shafts (bodies).

Figure 35. Callus union of two ribs due to fracture.

Figure 35.

Figure 36. Cervical rib (accessory rib).

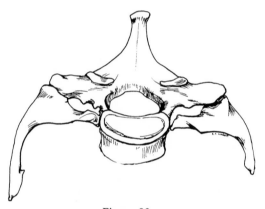

Figure 36.

Cervical ribs are enlarged costal elements (the bony bridges that span from the centrum to the rudimentary transverse process of cervical vertebrae) of the seventh cervical vertebra (Grant 1972). Such ribs may be rudimentary or fully developed and roughly similar in shape to normal first ribs but smaller and less "U" shaped. Radiologically, cervical ribs are found twice as often in females, 50 percent occur bilaterally, and in 5.6 per 1000 patients (Grant 1972). These ribs are, however, uncommon findings in archaeological samples. Cave 1941–42; Dwight 1887; Gladstone and Wakeley 1931–32; Hertslet and Keith 1896; Lucas 1915; Purves and Wedin 1950; Schaeffer 1942; Todd 1912.

SACRUM

Figure 37. Pathological conditions of the sacrum (posterior view).

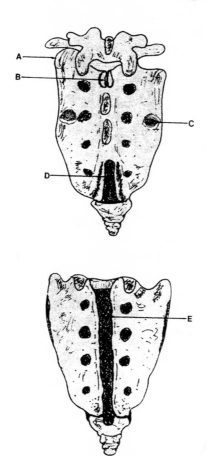

Figure 37.

A. Sacralization (sacralized lumbar vertebra, transitional lumbosacralization)—a congenital condition that results in partial or complete fusion of the most inferior lumbar vertebra (fifth or even sixth lumbar) to the sacrum. Occasionally a vertebra will exhibit characteristics of a lumbar vertebra on one side and sacral on the other resulting in the so-called hemilumbarization or hemisacralization (Bergman et al. 1988). Although a common clinical finding, these conditions are uncommon in most archaeological samples. (When computing stature, allowance must be made if there is an additional vertebra in the spine [Lundy 1988].)

B. Bifid spine (spina bifida, cleft spine, incomplete spina bifida)—
incomplete closure, fusion, or development of any of the neural arches
(spines). This condition may affect any vertebrae including those of the
sacrum (very common) and, although rare, may have resulted in paralysis.
Uncommon to common finding.

C. Accessory sacral facet (Derry 1911)—frequently, bilateral accessory
facets are present at the level of the first or second dorsal sacral foramen
for articulation with the ilium. The contact areas in the ilia are usually
located in the most inferior portion posterior to the auricular surface. To
detect these facets articulate the sacrum and ilium. Common finding.

D. Normal sacral hiatus—a canal that extends up one or two segments
from the most inferior segment of the sacrum. If the hiatus extends as far
as the third sacral segment/third sacral foramen (starting with the most
inferior segment), it should be scored as spina bifida. There is no total
agreement about critical criteria among researchers.

E. Complete spina bifida (possibly reflects the more serious condition
spina bifida cystica or aperta)—complete lack of fusion or development
of the posterior neural arches (spines) of the sacrum. Even if there is
incomplete development of all the sacral arches, extreme caution should
be exercised in "diagnosing" the condition as cystica or aperta (in the
lower vertebrae, unless the lumbar are affected, spina bifida is unlikely
to have clinical significance). Rare finding. Dickel and Doran 1989;
Strassberg 1982.

INNOMINATE

Figure 38. Degenerative changes and normal traits of the innominate (os coxa).

A. Enthesophyte (Dieppe et al. 1986; Resnick and Niwayama 1988)—
macroscopically, enthesophytes (some researchers refer to these projections as osteophytes) appear as spike-like projections, spicules, spurs,
and ridges of irregular ossification where tendons and ligaments attach
(enthesis) to the bone. This new bone growth may accompany old age,
obesity, or repeated acute minor stress related to a particular motion or
activity. Frequent sites of involvement are the linea aspera, trochanteric
fossa, greater and lesser trochanters of the femur, iliac crest, ischial crest
and tuberosity, ischial spine, and obturator foramen in the innominate;
attachment of the Achilles tendon in the calcaneus, supinator crest of the

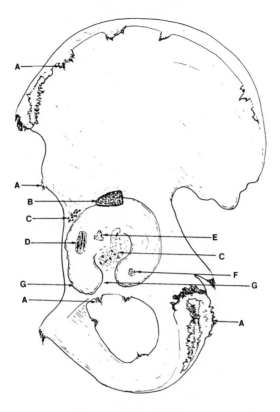

Figure 38.

ulna, radial tuberosity of the radius, and soleal line (popliteal muscle) of the tibia. In some cases, it is difficult to distinguish enthesopathy from normal skeletal variation (robustness). Enthesophytes, possibly reflecting DISH or fluorosis, are common findings in individuals 60+ years of age, and may be more prominent on one bone or one side of the body as a result of increased pulling stresses at these sites such as handedness or paralysis. Enthesophytes do not reflect osteoarthritis (inflammation of synovial joints).

B. Flange lesion (Knowles 1983; Wells 1976) or cotyloid bone (?) (Steele and Bramblett 1988)—the cause and correct name of this condition have not been definitely established. With some reservation, Wells (1976) and Knowles (1983) attribute a similar condition to acute, but temporary, dislocation of the femoral head on the rim of the acetabulum. This "lesion," present in one or both innominates, is approximately 2 to 3 cm in length and appears as an eroded area with either exposed trabeculae or as a smooth, flattened depression.

Earlier work by Terry (1933) describes a cotyloid bone as a large separate bone of the pubic portion of the acetabulum and resembling a bipartition (clearly not the same condition described by Knowles or Wells). However, there is an example of a cotyloid bone (labeled as such) fitting the description of the flange lesion in the Hamann-Todd collection in Cleveland, Ohio (Marc Kelley Pers. Comm. 1989). The latter example has the appearance of an accessory bone and is attached (postmortem) to the rim of the acetabulum with a wire.

Regarding etiology, an accessory bone along the rim of the acetabulum is more plausible, and similar in appearance to other accessory bones, than traumatic dislocation of the femur. Either flange lesions and cotyloid bones represent two different conditions (i.e., the former being traumatic and the latter developmental), or terminological confusion obscures the issue.

The authors have only encountered such "lesions" in two young individuals, both showing the porous form bilaterally. Care must be taken not to misinterpret normal irregularity of the acetabular rim with this condition. Examine the head and neck of the corresponding femur for signs of trauma or alteration reflecting OA. If encountered, it is probably best not to identify the condition by name but rather carefully describe the lesions. Rare finding.

C. Normal small pits above the acetabular rim and in the roughened central area of the acetabulum (but not on the articular surface).

D. Eburnation of the articular surface—eburnation results from degeneration and loss of joint cartilage. Continued joint movement results in bone on bone contact and sclerosis (thickening) of the subchondral bone (bone beneath the cartilage). Eburnated areas are ivorylike, polished (possibly yellow or brown in color), possibly grooved, and will reflect light (perhaps the best macroscopic method for detecting eburnation). Eburnation is a rare finding in young individuals but common in the elderly. Pauwels 1976.

E. Acetabular mark (normal variant)—a triangular-shaped defect or nearly detached u-shaped "tag" of bone located in the superior third of the acetabulum. This nonmetric trait may be a remnant of fusion of the bones forming the acetabulum (tri-radiate cartilage). Common finding. Anderson 1963; Saunders 1978.

F. Porosity (OA)—tiny holes or pits in any portion of the articular surface of the acetabulum. Common finding in the elderly and uncommon finding in younger adults (likely the result of trauma in young individuals).

G. Marginal osteophytes (OA)—common finding and one of the first indicators of OA of the acetabulum. To distinguish between marginal osteophytes and normal variation in the acetabular rim, feel the lip to detect if it is raised or sharp (osteophytes). In most cases the bony extension (osteophyte) will be very thin and clearly discernible as an extension of the acetabulum.

Figure 39. Pathological conditions of the pelvic girdle.

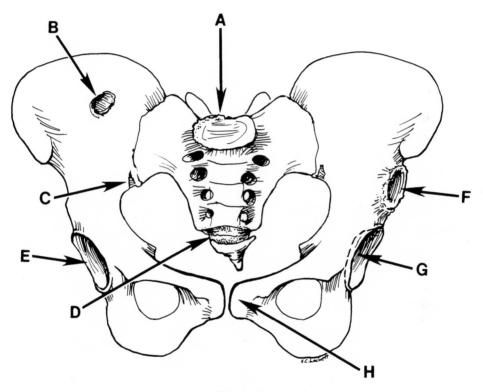

Figure 39.

A. Marginal osteophytes of the sacral promontory. Common finding accompanying increased age.

B. Psoas abscess—concavity or perforation in the inner surface of the ilium usually resulting from neoplasm, tubercular (tuberculosis), actino-mycotic or syphilitic invasion via the psoas muscle from the lumbar vertebrae (Simpson and McIntosh 1927). Rare finding in most skeletal populations. However, a skeletal sample with many cases of tuberculosis will show this trait in varying frequencies. Micozzi and Kelley 1985; Morse 1961, 1969; Steinbock 1976.

C. Preauricular groove/preauricular sulcus (groove of pregnancy, parturition groove)—a groove of varying width and depth anterior to the auricular surface of the ilium. Although a groove may be found in both sexes, those in females will usually be wider and deeper than in males. It is not possible to determine the exact number of births based on this groove and not all term pregnancies result in the formation of preauricular grooves. Common finding. Houghton 1974, 1975; Saunders 1978; Spring et al. 1989.

D. Fracture of the coccyx—in adults with fused coccygeal vertebrae there may be angulation and a visible fracture line or callus formation. Sometimes it is difficult to attribute the coccygeal curvature or lateral deviation to fracture because of normal variability in the coccyx. Uncommon finding.

E. Shallow acetabulum—comparison with other innominata will familiarize the observer with the normal variability of the acetabulum. This criterion for a shallow acetabulum is especially important in distinguishing traumatic from congenital hip dislocation. In congenital hip dislocation (Weinstein 1988) the acetabulum will be small and/or shallow, resulting in a poorly reinforced hip socket. If the acetabulum is of normal depth it is likely that the dislocation was due to acute trauma (see Figures 40 and 41). Rare finding.

Another condition to be aware of is an acetabulum that is too deep and results in a convex acetabular floor visible from the inner surface of the ilium (protrusio acetabuli. Alexander 1965; Wells 1976; Resnick and Niwayama 1981; Otto pelvis/intrapelvic protrusion, MacAusland and Mayo 1965). In protrusion acetabuli the greater trochanter of the femur may come into contact with the ilium during abduction (spreading of the legs). If this occurs there may be some destruction (wear facets) of the greater trochanter and ilium. Rare to uncommon finding.

F. Secondary acetabulum (false acetabulum, pseudarthrosis, pseudoarthrosis)—a secondary joint often accompanied with eburnation formed when the head of the femur dislocates from the acetabulum (Epstein 1973); the femur head may displace in any direction on the innominate (see Figure 41). At the site where the head comes to rest a new joint will form. The new joint will be rough and irregular and holds the femur in place by the formation of a circular "joint" composed of hypertrophic bone. Although the length of the affected leg may be reduced and its mobility restricted, the individual may have maintained a great deal of mobility. Rare to uncommon finding.

G. Normal acetabular depth.

H. Dorsal pitting (parturition pits, birthing scars) — circular or linear depressions or grooves on the dorsal surface of the pubic symphyses (not shown). These pits and/or grooves may resorb in old age. Although questionable, some researchers state that the number of births can be determined by examination of these pits and grooves. Common finding in females. Stewart 1957; Suchey et al. 1979.

Figure 40. Severe osteoarthritis of the left acetabulum (inset shows remodeled acetabulum and pseudarthrosis).

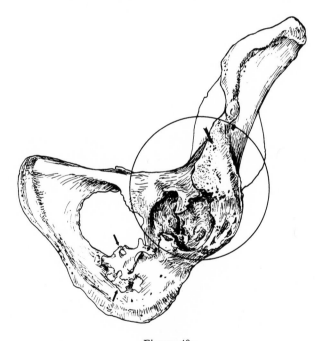

Figure 40.

This example represents chronic traumatic dislocation of the femur head in a 70-year-old white female. Note the flattened, pitted, and raised plateau of bone (large arrow) where the femur head came to rest. Also present are osteophytes and pitting in and around the acetabulum. The bony projections along the ascending ramus of the ischium (small arrows) represent fracture callus.

Figure 41. Common positions of the dislocated femur head.

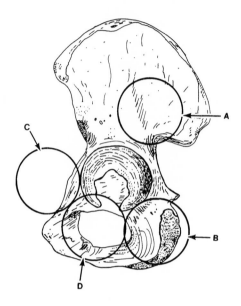

Figure 41.

A. Posterior superior.
B. Posterior inferior.
C. Anterior superior.
D. Anterior inferior.

Figure 42. Close up of preauricular sulcus/groove (internal surface of the right innominate).

The preauricular sulcus (Zaaijer 1866) is often attributed to be the result of pulling stresses of the anterior sacroiliac ligaments during birth delivery. Many researchers distinguish two types of preauricular grooves— the groove of pregnancy and the groove for ligamentous attachment (Houghton 1974, 1975; Saunders 1978); both are located anterior to the auricular surface of the ilium.

Some researchers attribute the groove of pregnancy to pulling stresses of the ventral sacroiliac ligament and subsequent inflammation (bleeding) associated with childbirth. Hormonal responses result in a loosening of the sacroiliac joint and localized growth resulting in a larger birth canal.

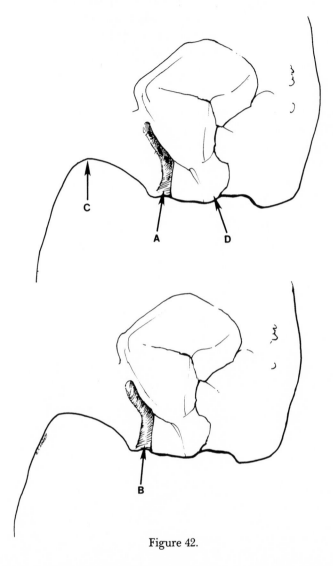

Figure 42.

A groove of pregnancy differs from a groove of ligamentous attachment in that the former will generally show discrete or coalesced pits or craters within the groove (which extends more superiorly along the anterior border of the auricular surface of the ilium). The latter will simply be a shallow, short groove present in both sexes (normal imprint of a strong ligament. Houghton 1975). Both types are common findings.

Spring and colleagues (1989) conducted a radiographic examination (hospital) of the preauricular groove in 190 females and 110 males (all adult). Deep grooves were noted in 4 of 41 nulliparous women and in 25 of 149 women with positive pregnancy histories. Additionally, no radiographic changes were noted before or after term pregnancies in six women. The authors concluded that " . . . the presence of a deep, radiographic preauricular sulcus is not necessarily an indication of past pregnancy."

A. Preauricular sulcus—groove of pregnancy.

B. Preauricular sulcus—groove for ligamentous attachment (both sexes may exhibit this groove).

C. Greater sciatic notch.

D. Auricular surface of the ilium.

MANUBRIUM AND STERNUM

Figure 43. Dorsal view of manubrium and sternum.

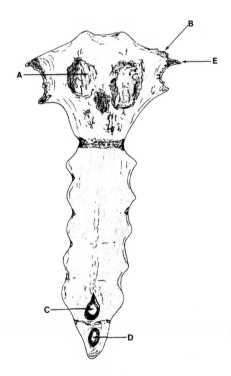

Figure 43.

A. Lytic lesions—deep cavitations in the manubrium may be the result of aneurysmal erosion stimulated by enlargement of the aorta or osteo-myelitis. Rarely, expansion (bulging) of the aorta may cause pressure erosion of the dorsal manubrium. Take care not to mistake normal irregularity and fissures in the posterior surface of the manubrium for disease. Rare finding.

B. Osteophyte—the normally smooth junction of the clavicle and manu-brium (sternoclavicular joint) will have small to large irregular projec-tions of bone. Common finding in the elderly.

C. Sternal perforation/aperture (cleft sternum, Knight and Morley 1936–37)—this developmental defect is often misidentified as a bullet wound or healed perforating wound of the sternum. This condition results from the lack of complete fusion of the lower two or three sternal segments (frequently between segments three and four) as they ossify separately from left and right centers (Grant 1972). Uncommon to rare (Saunders 1978) finding. Krogman 1940.

D. Xiphoid perforation—developmental defect resulting in a hole in the xiphoid process. Uncommon to common finding.

E. Ossification of the costal cartilage—the normally flexible costal carti-lage that serves to connect the ribs to the manubrium and sternum ossifies with age. In the elderly, the result is an irregular collar (tube-shaped) of bone especially common at the sternal end of the first rib. Common finding in the elderly.

CLAVICLE

Figure 44. Variation of the rhomboid fossa/pit and conoid joint (inferior view of the left clavicle).

The rhomboid fossa appears as a roughened, pitted, and oval depres-sion (Jit and Kaur 1986; Shauffer and Collins 1966) or a smooth and raised eminence (Williams and Warwick 1980) on the inferior surface of the clavicle for attachment of the costoclavicular ligament (Cave 1961) (sometimes referred to as the rhomboid ligament. Saunders 1978). Par-sons (1916) noted its presence in 10 percent of 183 clavicles. An examina-tion of 10,000 chest fluorograms of individuals between 8 and 70 years revealed this trait in 5 percent of the sample Khazhinskaya and Ginzburg 1975). A preliminary study (contemporary autopsy sample of 350 adults by one of the authors [RWM]) revealed that only males exhibit a rhom-

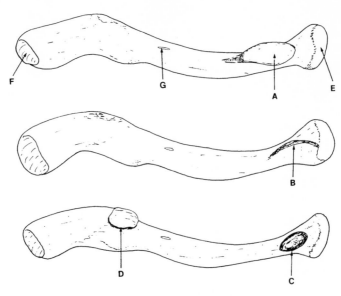

Figure 44.

boid fossa (C) longer than 15 mm. Although females commonly show this trait, it is usually much smaller than those in males. Etiology of the rhomboid fossa (commonly misdiagnosed as a lesion on radiographs) is unknown but it may be aggravated by strenuous activities of the pectoral girdle.

A. Raised plateau-like attachment site for the costoclavicular ligament. Uncommon finding.

B. Raised ridge-like attachment site. Uncommon finding.

C. Depressed crater-like fossa—the typical shape is oval or oblong. The fossa will show sharply defined cortical margins with a porous (trabecular) center. Common finding.

D. Conoid joint/process (Cockshott 1958)—a rare finding in most skeletal populations. The raised plateau-like bony extension articulates with the superior surface of the coracoid process of the scapula. Look for a corresponding facet on the coracoid. May be bilateral.

E. Sternal end.

F. Acromial end.

G. Normal nutrient foramen (may be multiple).

SCAPULA

Figure 45. Pathological and normal conditions of the scapula (lateral view).

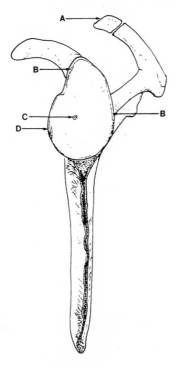

Figure 45.

A. Os acromiale (unfused acromial epiphysis, bipartite acromion, meta-acromion)—a nonmetric trait representing an ununited ossification center of the acromion (there are usually three separate centers for the acromion [Mudge et al. 1984]). The acromial epiphyses fuse between 16 and 25 years; only unfused epiphyses in individuals over 25 years of age should be scored as os acromiale (Saunders 1978). In a radiographic study of 1800 shoulders, Liberson (1937) found the trait bilateral in 62 percent of the cases and an incidence of 1.4 percent. Mudge et al. (1984) further reported that a possible correlation may exist between rotator cuff tears and os acromiale. Uncommon finding in most populations. Angel 1987; Chung and Nissenbaum 1975; McKern and Stewart 1957; Saunders 1978; Symington 1899, 1900.

B. Marginal osteophytes (OA)—raised, sharp margins along the glenoid fossa. Common finding accompanying middle age and older as well as individuals functionally stressing their shoulders in various activities (OA).

C. Surface porosity—small to large pits in the articular surface of the glenoid fossa. The normal fossa should be smooth and slightly concave or have an undulating surface. Common finding.

D. Repeated dislocation (subluxation/luxation) of the humeral head (glenohumeral joint)—this condition may result in flattening, erosion, porosity, and eburnation of the glenoid fossa. Although the condition of dislocation (clinically) may be fairly common in many populations, it is uncommon to rare to find discernible bony changes related to dislocation in either the humeral head or glenoid fossa.

Figure 46. Pathological conditions of the scapula (ventral and dorsal views).

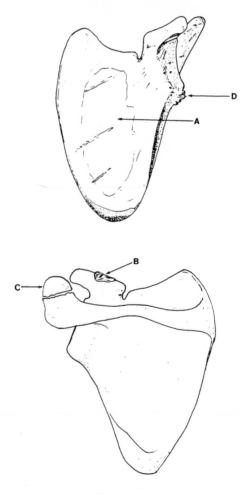

Figure 46.

A. Atrophy of the body—in elderly individuals, the scapular body may exhibit pleating, wrinkling, and perforation of the surface. Common finding in the elderly. Graves 1922.

B. Conoid joint—in some individuals the inferior surface of the distal clavicle will exhibit a raised, circular plateau of bone for articulation with the superior portion of the coracoid process of the scapula. Uncommon finding.

C. Os acromiale—familial trait resulting in an ununited acromial tip due to a failure of ossification. Uncommon to common finding.

D. Enthesophytes—irregular bony growths immediately below the glenoid fossa that result from functional stresses of the shoulder (insertion of the triceps muscle). The frequency of this condition may increase with old age.

HUMERUS

Figure 47. Pathological conditions of the humerus.

A. Surface (joint) osteophyte (OA). Common finding.

B. Flattening due to repeated dislocation, frequently inferiorly, of the humeral head from the glenoid fossa (Hill-Sachs disease. Danzig et al. 1980). Uncommon finding.

C. Marginal osteophytes (OA)—a distinct build up of bone encircling the head at its junction with the neck of the shaft. Common finding.

D. Surface porosity (OA)—be careful not to confuse this with postmortem alteration/destruction. Common finding in the elderly.

E. Normal foramina.

F. Pectoralis major cortical defect—a porous groove (cortical defect. Brower 1977; Caffey 1972; Keats 1973) for attachment of the pectoralis major muscle. In subadults this groove is probably a normal anatomical variant that remodels (fills in), often leaving a shallow depression in the adult. Cortical defects are common in subadults but rare in adults (Mann and Murphy 1989; Ann Stirland Pers. Comm. 1989. (The pectoralis groove is always lateral to the smaller teres groove.)

G. Teres major cortical defect—a porous groove similar in appearance to the pectoralis defect noted above. This groove is a normal feature frequently seen in subadults that normally remodels by adulthood but may be accentuated (deepened) due to acute or chronic trauma to the shoulder (e.g. prolonged tennis activity). Uncommon to rare finding in adults. Mann and Murphy 1989; Ann Stirland Pers. Comm. 1989.

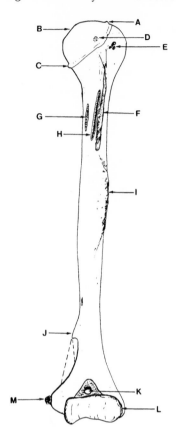

Figure 47.

H. Attachment site of the latissimus dorsi muscle—may be a groove in subadults (rarely) and young adults or a roughened and raised area in adults. Common finding.

I. Normally roughened and raised area (plateau-like) for attachment of the deltoid muscle.

J. Supra-condyloid process (Struthers 1873), supracondyloid process (Schaeffer 1942), supratrochlear spur (Saunders 1978)—a small, roughly triangular and hook-shaped exostosis projecting 5 to 7 centimeters above the medial epicondyle and varying in length from 2 to 20 millimeters (Genner 1959). This exostosis serves as an accessory ligamentous attachment site (dashed line) for origin of the pronator teres muscle. Through the tunnel formed by this fibrous band (ligament of Struthers) passes the median nerve and brachial artery. Congenital trait reportedly found in 7 of 1000 living subjects by Schaeffer (1942) and in approximately 1 percent of people of European ancestry (Barnard and McCoy 1946; Gray 1948; Terry 1921, 1926, 1930). This trait has a high rate of heritability and has been found in embryos (Adams 1934), children of all ages, and adults. Uncommon finding in most archaeological samples. Cady 1921; Dwight 1904; Hrdlicka 1923; Parkinson 1954; Pieper 1925; Rau and Sivasubrahmanian 1931; Witt 1950.

K. Septal aperture—a hole present in the olecranon fossa that may range from the size of a pinpoint to a large perforation. Although the etiology of this perforation is uncertain (e.g. congenital, developmental/ mechanical or hereditary), it is found more frequently in females than males and occurs in 4 to 13 percent of individuals (Bergman et al. 1988). Common finding. Bass 1987; Trotter 1934.

L. Marginal osteophytes. Common finding.

M. Enthesophyte of the medial epicondyle—bony growth due to pulling stresses of the pronator teres muscle (tendon attachment site). Uncommon to rare finding.

Figure 48. Pathological conditions of the right distal humerus (anterior view, actual size).

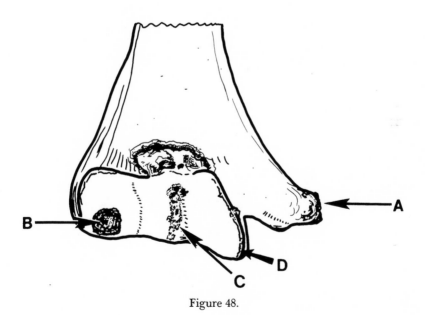

Figure 48.

A. Enthesophyte—hypertrophic development of the bone of the medial epicondyle. May be age (elderly) or activity (stress) related. Dieppe et al. 1986.
B. Osteochondritis dissecans (OD)—a developing lesion appears as a roughly circular, concave defect with steeply sloping edges (classic site is the medial femoral condyle). Most cases of OD result from trauma to a joint that causes destruction of the hyaline cartilage (covers all the long bone joint surfaces), and subsequent loss of the cartilage and underlying bone (such lesions may be confused with osteoarthritis). In some instances the concavity, which may affect any joint, may fill in with new bone (late lesion). Uncommon to rare finding in most populations. Barrie 1987; Clanton and DeLee 1982; Griffiths 1981; Lichtenstein 1970; Wells 1974.
C. Surface osteophyte—classic site of incipient (early-stage) osteoarthritis (OA) of the distal humerus. Porosity may be interspersed in or along this thin, raised ridge of new bone. Common finding. Ortner 1968.
D. Marginal osteophytes—the margin will be raised, sharp and irregular at any point along its rim. The development of a rim can best be detected by either rubbing your finger along the "sharpened" rim or catching the lip with your fingernail. The presence of a slightly raised rim indicates early or minimal OA and likely presented no pain, discomfort or restriction of the joint. This is a common finding in populations that use their forearms a lot (e.g. Easter Islanders who carved large monoliths), as well as in the elderly. Common finding.

Figure 49. Pathological conditions of the left distal humerus (posterior view, actual size).

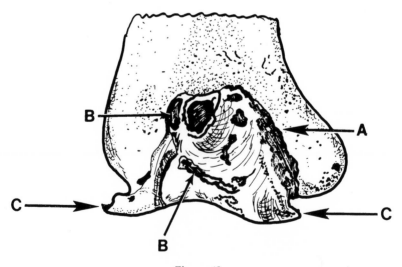

Figure 49.

Severe OA of the posterior distal left humerus—note the large osteo-phytes (A) along the medial margin of the articular surfaces ("mush-rooming") and extension into the olecranon fossa. Severe forms of OA of the distal humerus may also result in bony deposits (hyperplasia resem-bling small mounds), large pits (B), and distortion of the articular surface (C). Although similar severe changes may be the result of local acute trauma, most instances are associated with old age. Look for frac-tures of the proximal ulna and distal humerus. The severe form is uncommon.

RADIUS AND ULNA

Figure 50. Pathological conditions of the right radius (lateral view).

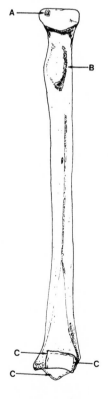

Figure 50.

A. Surface porosity (OA). Common finding.
B. Enthesophyte—a ridge-like build up of bone along the lateral border of the radial tuberosity (biceps muscle attachment). This enlargement

may be age and activity related that increases due to stresses of the arm. Uncommon to common finding.

C. Marginal osteophytes of the distal joint surface—sharp, raised, and irregular rim. Common finding.

Figure 51. Typical bony responses to mechanical stress at tendon and ligament attachments (enthesophytes), osteoarthritis (OA), rheumatoid arthritis (RA), and Colles' fracture.

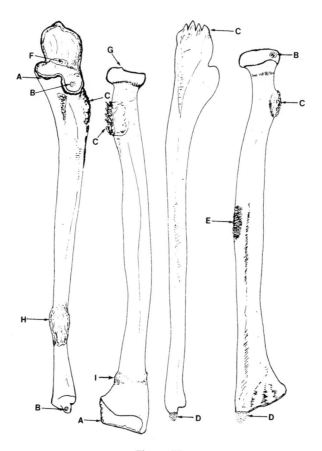

Figure 51.

A. Marginal osteophytes—sharp, well-defined and raised areas of bone along the articular margin (OA). Common finding, especially in the elderly.

B. Surface porosity—pits anywhere in the articular surface (OA), possibly accompanied by osteophytes. Common finding, especially in the elderly.

C. Enthesophyte—bony development of the biceps muscle insertion (radius), triceps (olecranon process of the ulna), and supinator (supinator crest of the medial ulna). Common finding in populations that use their arms in strenuous activities. Enthesophyte development (enthesitis, enthesophytosis) may increase in size in the elderly as the result of many years' wear and tear (pulling) of these tendon (muscle) and ligament sites.

D. Erosion of the styloid process of the ulna and radius—one of the criteria associated with rheumatoid arthritis (Resnick and Niwayama 1988). Osteoarthritis (in old age) may resemble this condition. Consult a rheumatologist if there is erosion of the ulnar or radial styloid processes with little or no bony proliferation. Other indicators of RA are eroded odontoid process of the axis (dens epistropheus), involvement of the hands, feet, and rarely, the spine (Lonstein and Hochschuler 1989). Distinguishing RA from a number of other conditions is extremely difficult and requires the assistance of a rheumatologist. Fracture of the styloid process is common in Colles' fracture. Dieppe 1986.

E. Normal roughened area for attachment of the pronator teres muscle.

F. Trochlear notch (notching)—a raised, narrow band of bone of varying size extending across and possibly dividing the articular surface of the coronoid fossa in two. There is some dispute whether this trait, in varying forms, may be a normal finding in all individuals or a nonmetric trait (Saunders 1978). Common finding.

G. Surface osteophyte—a small raised area of bone anywhere on the articular surface. Common finding in the elderly reflecting OA.

H. Parry fracture ("nightstick," defense fracture of the forearm)—results when a person raises his or her arm to prevent being struck in the face or head (the ulna is the usual site of fracture). Uncommon to common finding depending on the amount of warfare or domestic violence in a population. Knowles 1983.

I. Colles' fracture—fracture of the distal third of the shaft resulting from breaking a fall on an outstretched hand (the force of the fall is directed through the radius). Although first described in 1814 by Abraham Colles referring to fracture of both the distal radius and styloid process of the ulna (Griffiths 1981), most researchers now identify any fracture of the distal radius (shaft) as a Colles'. Common finding in contemporary elderly individuals (McAuliffe et al. 1987), uncommon finding in most archaeological samples. Squire 1964; Wilson 1930.

FEMUR

Figure 52. Pathological changes of the head and neck of the femur (anterior view, left femur).

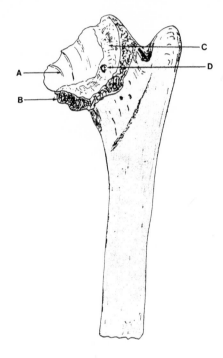

Figure 52.

A. Severe eburnation—ivory-like grooves and flattened areas in the articular surface (OA). Eburnated bone can range in size from extremely small (early stage) to areas involving the entire articular surface. Eburnation can be detected by its smooth shiny appearance that reflects light. Eburnated bone frequently yellows and resembles old piano keys. Uncommon to rare finding associated with the elderly.

B. Periarticular bone (hypertrophic bone, periosteal osteophytes, "mushrooming")—loss of bone height in the femur head has resulted in new bone growth near and encircling the neck (OA). Uncommon to common finding associated with the elderly.

C. Flattened areas of bone where there has been degeneration and resorption of most of the femur head (OA). Uncommon to common finding associated with the elderly. Uncommon to rare in many archaeological samples.

D. Bone cyst (subchondral cyst) below the articular surface—common radiographic finding in elderly individuals with osteoarthritis and avascular/ischemic necrosis (dead bone usually resulting from trauma and loss of blood supply to the femur head; especially common in very old individuals; Claffey 1960; Jaffe 1969, 1975). Etiology unknown. Dieppe et al. 1986.

Figure 53. Osteoarthritis and enthesopathy of the proximal femur (posterior view).

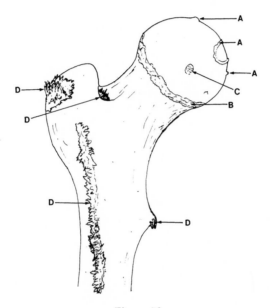

Figure 53.

A. Surface osteophytes.
B. Marginal osteophytes—raised lip of bone along the junction of the head and neck of the femur.
C. Surface porosity—small to large pits in the articular surface.
D. Enthesophytes—new bone at the insertions of tendon and ligaments. Enthesophytes appear as irregularly raised areas (excrescences), roughened attachment sites, bony spicules, spurs or projections commonly found on the greater and lesser trochanters, trochanteric fossa, and linea aspera. This condition is due to stresses at tendon and ligament attachments (any) to bone. Some researchers state that enthesophytes occur only at the site of tendon and ligament insertions (Dutour 1986) because more stress is placed on the insertions, rather than the origins, due to the smaller area of

attachment fibers into the bone. Enthesophytes are strongly correlated with old age. Care must be exercised not to confuse enthesophytes with myositis ossificans. Dutour 1986; Merbs 1983; Resnick and Niwayama 1988.

Figure 54. Tuberculous destruction of the femur head (anterior view, left femur; inset shows detail of the remodeled head and neck).

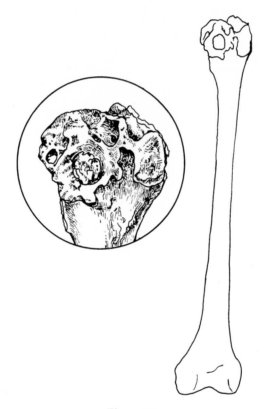

Figure 54.

Figure 55. Severe osteoarthritis of the femur head.

A. Medial view—the large, flattened area of bone posterior to the femur head (arrow) articulated with the acetabulum.
B. Posterior view—note the irregularly-shaped head ("mushroomed"), shortened neck, and macroporosity. This is the typical appearance of severe osteoarthritis of the proximal femur and may represent either Perthe's disease or slipped capital femoral epiphysis. In a young individual, Perthe's or slipped capital femoral epiphysis should be suspected.

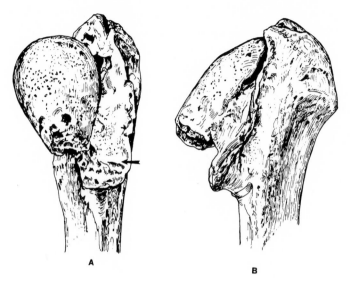

A   B

Figure 55.

*Figure 56. Cross-section of the proximal femur demonstrating common
indicators of osteoarthritis. Hough and Sokoloff 1989.*

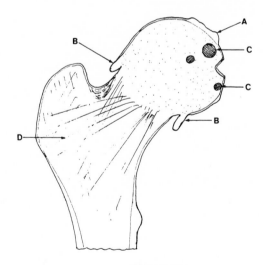

Figure 56.

A. Surface osteophytes. Common finding in the elderly.
B. Marginal osteophytes. Common finding in the elderly.
C. Subchondral bone cysts—x-rays will reveal radiolucent (dark) areas

where the cysts exist. Common finding accompanying osteoarthritis, rheumatoid arthritis, and avascular necrosis (Resnick and Niwayama 1988).

D. Trabeculae (spongiosa)—the normal internal bony architecture of the cortex.

Figure 57. Anterior view of proximal femur showing a cervical fossa of Allen.

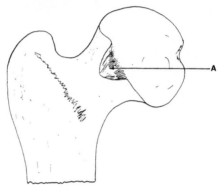

Figure 57.

This facet (arrow) is the result of contact of the tendon of rectus femoris on the neck of the femur (Kate 1963), not contact with the acetabulum (the so-called anterior acetabular imprint. Meyer 1924). It has further been suggested by Kate that this fossa is the evolutionary result of the erect posture and is not due to squatting. There are a variety of depressions, raised plaque-like areas, and cribiform (porous/trabecular) lesions that may be present at various locations on the femur neck. Allen's (1882) original description of the femur neck stated that it "is marked in front near the articular surface by a faint depression, which is often cribiform in appearance and may receive the name cervical fossa." Findings by subsequent researchers have since made it difficult to distinguish one facet from another with the addition of such terms as the imprint of Bertaux (1891), Testut (1911), and Poirier's facet (1911). Common finding. Finnegan and Faust 1974; Kostick 1961, 1963; Meyer 1924; Saunders 1978.

Figure 58. Posterior view of the proximal femur showing a Walmsley's (1915) facet.

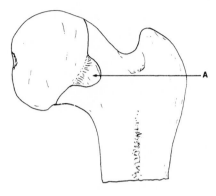

Figure 58.

This facet (arrow) appears as a smooth, depressed, and well-defined extension of the posterior articular surface of the femur neck. In some of the older literature Walmsley's facets may be referred to as posterior acetabular imprints (Meyer 1924). Common to uncommon nonmetric trait.

Figure 59. Legg-Calve'-Perthes disease (Perthes' disease. Karpini et al. 1986) of the femur (posterior view, left femur).

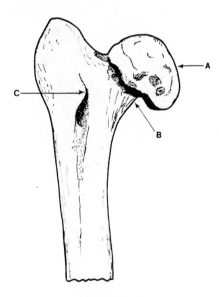

Figure 59.

The example used here is Perthes' disease with premature closure of the proximal growth plate (physis) and severe deformity of the left femur head in a 14-year-old male. Similar deformity could, however, be the result of congenital dislocation of the hip, slipped capital femoral epiphysis (Busch and Morrissy 1987; Schlesinger and Waugh 1987; Schoenecker 1985), or trauma with resulting arthritis. Check the femoral neck length—if normal, probably Perthes' and if short, more likely slipped epiphysis.

Note the flattened, distorted (mushroomed) head and absence of the fovea capitis (for ligamentum teres attachment). The head is displaced inferiorly and the neck (B) is short in comparison to the normal right femur. The diameter of the left femur head is much larger than that of the right. Note the position of the head in relation to the tip of the greater trochanter—the head is much lower (trochanter overgrowth). The epiphysis of the right femur is incompletely fused (entire symphyseal line is visible) while that of the left is completely fused (coxa brevis. Bowen et al. 1986).

The left acetabulum (not shown) is also distorted, flattened, and exhibits an irregular (wavy) articular surface; this form is rarely seen in subadults. Old adults, however, may exhibit distorted femur heads due to trauma and avascular necrosis. Although the etiology is much debated, Perthes' disease likely results from circulatory disturbances of the epiphysis (Schwarz 1986) although trauma, infection, malnutrition, and endocrine disturbances have also been suggested. Caterall 1982; Jaffe 1975; Thompson and Salter 1986.

Figure 60. Fractured femur neck (anterior view; arrow denotes the fracture).

A common occurrence in the elderly ("broken hip"), especially postmenopausal and senescent females. Fractures of the neck can result from acute trauma such as a fall or in the aged, an act as simple as sitting down or lifting oneself from a chair. Fractured femur necks often lead to avascular necrosis (bone death) and death of the individual due to disability and complications in the elderly (e.g. pneumonia). In contemporary populations, 10–20 percent of elderly people who sustain "hip" fractures die within 6 months of injury (Genant 1989). Cummings et al. 1985; Gardner 1965.

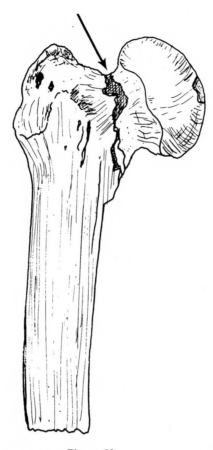

Figure 60.

Figure 61. Myositis ossificans traumatica of the proximal right femur.
Note the exuberant bone formation (often referred to as an exostosis) which would have extended into the muscles of the upper thigh. Myositis ossificans traumatica results from trauma to the muscle and subsequent ossification.

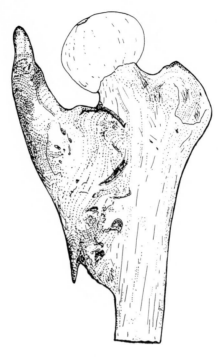

Figure 61.

Figure 62. Degrees of severity of osteoarthritis of the distal femur (anterior view).

The darkened areas at the margin of the articular surface represent irregular, raised osteophytes. In the early stage of OA, osteophytes appear as small, sharply defined and raised areas detectable either with the unaided eye or by touch (A). (Your fingernail will catch on this ridge of new bone.) Gradually new bone will cover more and more of the distal femur as it encroaches on the joint surface.

When distinguishing surface from marginal osteophytes look for altered joint surfaces and "bumps" (surface osteophytes). If the osteophytes extend onto the joint surface they should be scored as surface osteophytes. If both the margin has been extended by new bone growth away from the joint surface and the joint surface itself is affected, score both conditions (joint and margin osteophytes) as present. This criteria can be applied to any joint in the skeleton. Be consistent in distinguishing slight from moderate and severe. Remember, what might be a severe form of OA (C) in one population may in comparison only be moderate (B) in another.

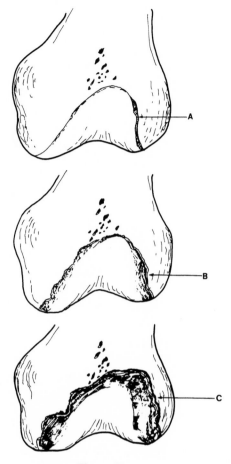

Figure 62.

A. Slight marginal osteophytes.
B. Moderate marginal osteophytes.
C. Severe marginal osteophytes.

Figure 63. Pathological and normal conditions of the distal femur (anterior view).

A. Marginal osteophytes—raised, sharp lipping indicative of OA that may be the result of acute trauma (e.g. knee injury) or associated with old age. Common finding in the elderly.
B. Articular surface porosity—small to large pits accompanying old age. Common finding in the elderly.
C. Articular surface osteophyte—raised areas of bone on the articular surface. Common finding in the elderly.

Regional Atlas of Bone Disease

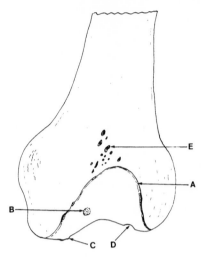

Figure 63.

D. Osteochondritis dissecans (OD) (Wells 1964, 1974), osteochondritis dessicans (Smillie 1960)—a scooped-out lesion, usually in the medial condyle of the femur (Figures 63 & 65). First described in 1870 by James Paget, OD is usually attributed to avascular necrosis that begins with acute trauma to the joint. However, numerous other causes have been suggested including abnormal ossification of the epiphyseal cartilage, contact with the tibial spines, hereditary influences, and generalized disorders.

The lesion begins with the separation of a portion of cartilage and underlying bone fragment resulting in a "loose body" in the joint, also called "joint mice." In time the bone and cartilage are resorbed resulting in a crater-like defect in the mature skeleton (Bradley and Dandy 1989). OD can affect any joint but has a predilection for the medial condyle of the femur appearing during the second decade of life. Hughston et al. (1984) found 78 defects in the medial femoral condyle and 17 defects in the lateral femoral condyle in a sample of 83 patients with OD. Clinically, OD is seen in 15 to 21 cases per 100,000 in the femur although many cases may go undetected unless accompanied by pain and detected on x-ray. Uncommon to common finding in most skeletal samples. Barrie 1987; Clanton and DeLee 1982; Griffiths 1981; Manchester 1983.

E. Normal nutrient vessel foramina.

Figure 64. Pathological and normal conditions of the distal femur (posterior view).

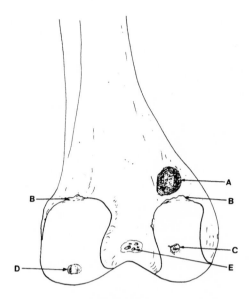

Figure 64.

A. Distal femoral cortical excavation (Resnick and Greenway 1982) (subperiosteal cortical defect, cortical defect) ("tendon lesion", Hrdlicka 1914)—a crater-like lesion located above the medial condyle at the attachment of the medial head of the gastrocnemius muscle (usually bilateral). There are many reasons given for this defect but no definitive conclusion as yet. Most researchers believe it is due to the repeated pulling stresses or acute trauma of the gastrocnemius muscle, hyperemia (increased blood velocity), and localized bone resorption. A common finding in subadults that is likely related to rapid bone remodeling during youth. Hrdlicka (1914) noted its presence in the skeletons of ancient Peruvian adolescents but not in children.
B. Marginal osteophytes (OA).
C. Surface osteophyte (OA)—raised area of bone on the articular surface.
D. Eburnation—a number of factors (e.g. congenital, developmental, and traumatic) may cause destruction and loss of cartilage resulting in areas of denuded bone and polishing (sclerosis) of the articular surface.
E. Normal nutrient vessel foramina.

Figure 65. Osteochondritis dissecans (OD): drawings of the stages and out-come of the condition (Reprinted with permission from Harry J. Griffiths, *Basic Bone Radiology,* Appleton-Century-Crofts, Norwalk, CT, 1981, as modified).

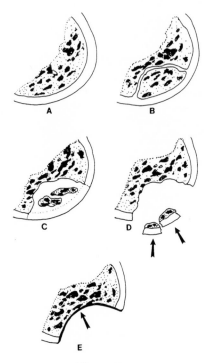

Figure 65.

The term osteochondritis is used to cover a large group of conditions (over 80 different types) that predominately affects overweight children between the ages of 5 and 11 years (Griffiths 1981). Uncommon finding in archaeological samples. For an interesting discussion of the etiology of OD see Barrie (1987).

A. Normal joint.

B. Infarcted area with separation of a large bone and cartilage fragment.

C. Resorption of the contents and showing future line of weakness in cartilage.

D. Formation of loose bodies ("joint mice"—arrows).

E. With time the defect (osteochondritic pit) heals and appears as a smooth-walled crater (arrow).

Figure 66. Left femur and patella (from below) showing bony changes due to prolonged flexion of the lower leg.

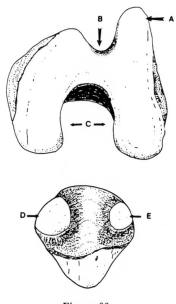

Figure 66.

A. Narrow and seemingly elevated lateral condyle relative to the medial condyle. The lateral condyle serves to guide and hold the quadriceps tendon in place.

B. Deep trochlear fossa (patellar surface) due to contact pressure erosion of the quadriceps tendon. The average depth of most adult trochleae is 5.2 mm but some may be as deep as 10 mm (Casscells 1979). Trochleae deeper than 10 mm may represent a pathological condition.

C. Wide intercondylar fossa.

D. Large, raised, and widely-spaced lateral articular facet.

E. Small, raised, and widely-spaced medial articular facet. If changes of this type are noted in either the femur or patella, it is almost certain that the individual endured many years with the lower legs in flexion (e.g. legs drawn toward the chest), possibly due to paralysis. Increased contact pressure of the quadriceps tendon against the distal femur results in erosion of the trochlear surfaces. A number of conditions including poliomyelitis, spinal cord trauma, and spina bifida, can cause flexion contractures of the legs. The finding of deep trochlear fossae or widely-spaced patellar facets holds cultural implications of caring for the individual who might have been unable to properly care for him/herself. Rare finding in most archaeological samples and uncommon finding in contemporary populations.

Figure 67. Cross-section of a femur with sclerosing osteomyelitis.

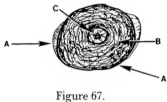

Figure 67.

Note the original cortex (dark outer line; A), and the new bone (B) that has nearly obliterated the medullary canal (C). Detection of endosteal new-bone formation can only be made by using one of the radiological methods (e.g. x-raying, nuclear magnetic resonance [NMR], or CAT scan) or destruction of the bone by coring or sectioning with a saw. Seek an experienced researcher if this condition is found or suspected. Uncommon to common finding.

Figure 68. Neuropathic joint (Charcot's joint, anterior view of knee).

Neuropathy may affect any joint but shows a predilection for the large weight-bearing joints such as the knee. Although the specific etiology is debated, it is likely that Charcot's joints (severe osteoarthritis) develop after sensory loss to a joint and continued joint stress (Hellman 1989). This example depicts a hypertrophic Charcot joint of the knee involving the femur, tibia, and patella. Note the massive exuberant new bone and misalignment of the femur and tibia (the midline axis has changed). Rare finding. Brower and Allman 1981; Eichenoltz 1966; Jaffe 1972; MacAusland and Mayo 1965; Ortner and Putschar 1985.
A. Dashed line represents the original position of the femur, tibia, and patella.
B. Solid line shows misalignment of the femur and tibia and enlargement of the patella.
C. Thin, fragile periarticular new bone formed around the knee joint. Other frequent findings in Charcot's joints are bony ankylosis, loss

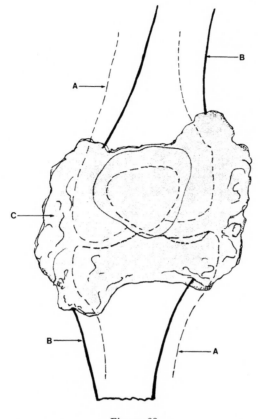

Figure 68.

of joint movement, and eburnation of the joint surfaces. The exuberant new bone is very brittle and may not survive recovery from the ground.

TIBIA

Figure 69. Amputated left tibia.

A. This tibia shows amputation at the midshaft with slight remodeling visible (a). Based on the amount of remodeling (woven bone) it is evident that only a short period of time, perhaps a few weeks, has elapsed since the inflammation and/or infection began. Normal bone (b).

B. This tibia shows a well-healed (years) amputation at the midshaft. Note the atrophy (pegging), healing, and loss of normal shape and circumference above the site of amputation (a). The medullary cavity has narrowed and closed due to remodeling and healing.

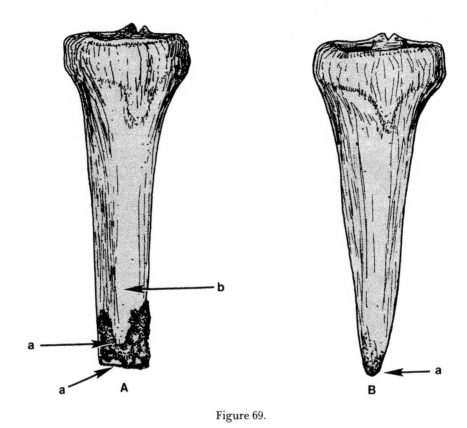

Figure 69.

Figure 70. *Actinomycosis of the tibia.*

Note the enlargement and deformity of the bone as well as the characteristic multiple scooped-out, cratered lesions (A) highly suggestive of
fungal infections (see Fungal Infections).

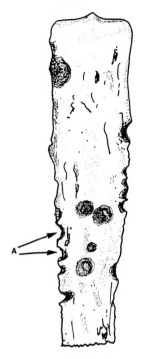

Figure 70.

Figure 71. Active periostitis (lateral view of right tibia).
A. Periostitis results from inflammation of the periosteum (tight outer sheath of all long bones and the skull) as the result of direct trauma such as a blow to the tibia (possibly resulting in an ossified periosteal hematoma. Day 1960), infection that has traveled to the tibia through the blood (e.g. syphilis), venous insufficiency as in varicose veins (Daniels and Nashel 1983; Gensburg, et al. 1988), scurvy, and a host of other factors. Common usage of the term periostitis generally refers to bone formed only on the outer cortex while osteitis refers to changes within the cortex, and osteomyelitis to changes that affect both the marrow and bone. Many times it is difficult, if not impossible, to ascertain the cause of the periosteal remodeling.

Periostitis, in the active stage, is easily recognized by its color, texture, and raised appearance. Active periostitis (A) is usually grayish in color, pitted (increased vascularity), striated, and has well-defined, raised margins (B). The best description that the present authors can offer of active periostitis is that it looks like small, loosely attached patches of tree bark.

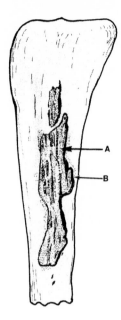

Figure 71.

Figure 72. Pathological conditions of the proximal tibia (posterior view).

A. Spiking of the tibial spines—early indicator of osteoarthritis (OA). The spines will be slightly elongated and sharp to the touch. Common finding in the elderly.

B. Surface osteophytes—raised areas (OA) on the articular surface. Common finding associated with increasing age.

C. Eburnation—polished joint surfaces due to bone on bone contact after loss of the cartilage (OA). May result from acute trauma, disease, or old age. Uncommon finding.

D. Marginal osteophytes—irregular, raised bony margins (OA) resulting from enchondral ossification.

E. Enthesophytes of the soleal (popliteal) line—this attachment site will be raised and irregular. If you can grasp this ridge and hold the bone by it, it should be scored as present (mild, moderate, or severe). Normal tibiae will exhibit a roughened soleal line, so the amount of development of the ridge will dictate the severity for scoring purposes. Some researchers believe that development of the soleal line is a pseudoperiostitis and not pathological.

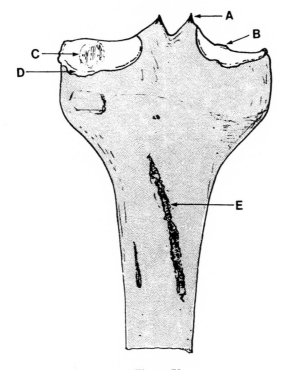

Figure 72.

Figure 73. Pathological conditions of the proximal tibia (anterior view).

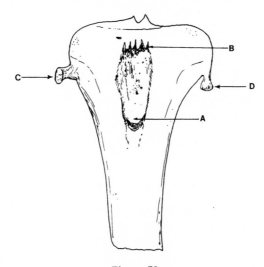

Figure 73.

A. Osgood-Schlatter's disease—traumatic fragmentation, inflammation (apophsysitis), and irregular ossification of the tibial tubercle in childhood. The inferior border of the tibial tubercle may partially "pull away" from the tibia and result in a gap that usually heals by adulthood. Although a common radiographic finding in contemporary populations, this condition is uncommon to rare in most archaeological samples. Dieppe et al. 1986; Resnick and Niwayama 1988; Wells 1968.

B. Enthesophytes—bony spicules or projections present in the antero-superior surface of the tibial tuberosity. The normal tuberosity will be smooth and show no projections at this site. Common finding in the elderly and individuals with DISH (does not represent OA).

C. Exostosis—rounded, smooth or irregular bony projections, usually extending perpendicular to the long axis of the diaphysis. Frequent sites of involvement are the metaphyseal areas of the shoulder, knee, and ankle (Spjut et al. 1971). Uncommon finding in most archaeological samples.

D. Osteochondroma (cartilaginous exostosis)—solitary or multiple cartilage capped projection (in the living) that typically extends parallel to the long axis of the diaphysis and away from the nearest joint. Solitary exostosis is the most common (90%) benign bone tumor (Lange et al. 1984). Usually located in the metaphysis. Uncommon finding in most skeletal samples. Spjut et al. 1971.

Figure 74. Pathological conditions of the distal tibia and fibula.

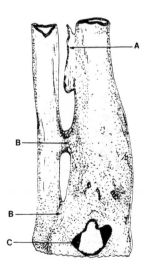

Figure 74.

A. Myositis ossificans (traumatica = of a traumatic origin)—trauma and resulting inflammation of muscle (fascia as well as tendon) may result in the ossification of muscle into bone. Any soft tissue trauma can result in the proliferation and transformation of tissue into bone (ectopic or heterotopic bone). If much of the skeleton is affected, the condition may reflect rare myositis ossificans progressiva (hereditary). If attached to bone, the condition almost inevitably reflects tendon or ligament involvement. Myositis ossificans is most frequently seen on the posterior surface of the tibia and femur. Uncommon to rare finding. Griffiths 1981; Resnick and Niwayama 1988; Schajowicz 1981.

B. Bony ankylosis (anchylosis, fusion)—many diseases as well as congenital factors or trauma (especially fracture) and subsequent remodeling may result in bony union of two or more bones. The bones fuse through the proliferation of callus, osteophytes or exostoses. Bony ankylosis of the joints of major long bones is uncommon to rare. When ankylosis of a joint is encountered, seek the advice of a specialist. However, ankylosis of the ribs to the vertebrae (AS) and bones of the hands and feet, especially the middle and distal phalanges of the feet is an uncommon to common finding.

C. Cloaca (draining sinus, fistula)—cloacae are frequent findings in chronic suppurative osteomyelitis. The light island of bone visible within the cloaca (in the drawing) is a sequestrum, that is, an isolated dead bone fragment that became walled off by exudate, granulation tissue, or scar. The outer envelope of irregular bone is an involucrum (i.e., sheath of newly-formed bone). As stated by Carney and Wilson (1975), "The special pathologic tissue reactions which are characteristic of suppurative osteomyelitis are the lysis of bone, the formation of new bone, and the presence of dead bone." If the fragment(s) is small enough it will gradually be resorbed (absorbed). In dry bone the sequestrum may be freely mobile and rattles when shaken. Uncommon to rare finding in archaeological specimens. May be a common finding in hospital and battle-related skeletal samples where surgical intervention prolonged and often exacerbated infection (e.g. Civil War soldiers).

Figure 75. Selected pathological conditions of the tibia (left, anterior view).

A. Chondroblastoma—generally benign tumor originating in cartilage and often found near the epiphysis (Spjut et al. 1971). Chondroblastomas affect males 2 to 1 over females. Periosteal reactions are rare. Rare finding.

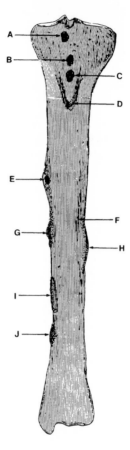

Figure 75.

B. Enchondroma—benign tumor originating in cartilage and located within the medullary cavity (Dahlin 1967). Solitary enchondromas are most often found in the small bones of the hands and feet with the long bones much less frequently affected. Uncommon finding.

C. Osteoblastoma—solitary, benign bone tumor that most often affects the vertebrae and long bones of the extremities (Spjut et al. 1971). This tumor is often misinterpreted as osteoid osteoma, aneurysmal bone cyst or chondrosarcoma (Pochaczevsky et al. 1960). Uncommon finding.

D. Osgood-Schlatter's disease—traumatic, partial avulsion and inflammation (apophysitis) of the tibial tubercle during childhood (not to be confused with complete avulsion of the tibial tubercle). Although a common finding in contemporary children, the condition is difficult to detect in adults. Uncommon to rare finding. Wells 1968.

E. Osteoid osteoma—a benign tumor that appears as a central lucent nidus (dark defect radiographically) with dense (sclerotic) surrounding bone. (Not to be confused with a Brodie's or tuberculous abscess or sequestration in chronic osteomyelitis [Mahboubi 1986].)

F. Linear striations—although healthy tibiae and femora exhibit linear striations running parallel to the long axis of the diaphysis, there may be an increase in these striations as well as tiny pits accompanying periosteal remodeling (periostitis). Look for shallow, parallel grooves near healed trauma, especially in the tibia, femur, and fibula in that order. Common finding.

G/H/I—Healed stress fracture/ossified subperiosteal hematoma/active periostitis—these forms of trauma are often difficult to differentiate from one another. The former results from a minor fracture and callus formation, accompanied by a lifting of the periosteum and periosteal new bone formation. Ossified subperiosteal hematoma (Kullmann and Wouters 1972) results from local acute trauma such as a kick to the shin causing a hematoma (blood clot). Periostitis results from periosteal new bone formation when the thin periosteum is lifted. Lifting of the periosteum may be the result of many conditions, including syphilis, venous insufficiency (e.g. varicose veins or blocked arteries), scurvy, and other causes.

Periostitis is a common condition in most archaeological populations, with the tibia being most often affected. Active periostitis has the appearance of tree bark that has been applied to the surface of the bone. The margins of an active lesion will be slightly elevated and well defined. The active new bone will also be of a different color than the surrounding bone. It should be remembered that the periosteum in fetuses, newborns, and children is only loosely adhering to the cortex and can be easily lifted resulting in massive periosteal remodeling (e.g. congenital scurvy or syphilis. Toohey 1985).

J. Healed periostitis—this condition may be the result of G, H, or I noted above. However, healed periostitis will show ill-defined and sloping margins. The raised area (trauma site) will be the same color and texture as the surrounding bone (e.g. smooth and possibly pitted).

Only a few tumors are discussed because of the extreme difficulty in diagnosing these lesions in dry bone. For further information refer to Chapter IX.

Figure 76. Sagittal section of a sabre-shin tibia ("boomerang leg").

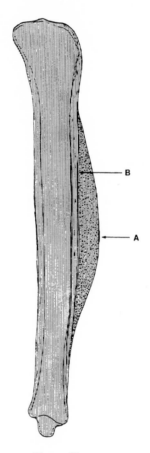

Figure 76.

Tibiae affected by acquired or congenital syphilis or some of the other treponematoses (e.g. yaws in tropical areas) may exhibit periostitis (A) along the anterior and middle portion of the shaft, resulting in an anterior enlargement ("bowing") of the bone. Note the underlying original cortex (B). Sabre-shin tibiae were noted in 9 of 100 (Caribbean) and 11 of 271 (native-born Americans in Boston) patients with late congenital syphilis (Fiumara and Lessell 1983 and 1970 respectively). Hackett 1976; Krogman 1940.

Figure 77. Pathological lesions of a tibia exhibiting chronic osteomyelitis.

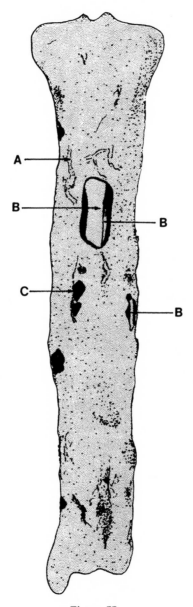

Figure 77.

A. "Snail track"—typical finding in chronic suppurative (pus forming) specific or nonspecific osteomyelitis. "Snail tracks" are smooth, shallow grooves in the outer cortex. Small to large pits (vascular foramina) may be present in the surface of these grooves. The undulations, large depressions, and furrows in the surface of the tibia would likely have been filled with granulation tissue (pus between the periosteum and cortex).

In those cases where there is a general thickening of the cortex but no cloaca, nonsuppurative (sclerosing) osteomyelitis and syphilitic osteomyelitis should also be considered. Proliferative periostitis and periostitis ossificans are descriptive terms that can be used for bones that are grossly thickened along much of their shaft. If the tibiae are greatly thickened, consult a paleopathologist or other expert for assistance with a possible differential diagnosis (e.g. possibly tuberculosis or syphilis). Uncommon to common finding depending on the population under study.

B. Sequestrum (necrotic bone)—dead fragments of bone (sequestra = plural) that have lost their nourishment supplied by the blood (avascular, devitalized). These bone fragments, if small, become detached and ultimately resorbed through the healing process. Rare finding in most populations. There are numerous accounts of soldiers wounded in the Civil War who pulled small to large sequestra from draining sinuses in the skin years after injury. These bone fragments were often sent to the Army Medical Museum in Washington, D.C. for documentation and inclusion in the soldier's medical and surgical record.

C. Cloaca (hole)—draining sinus for pus. Uncommon to common finding in historic skeletons and rare in American Indians from the Plains. Carney and Wilson 1975; Hertzler 1931; Ortner and Putschar 1985.

PATELLA

Figure 78. Pathological, anomalous, and normal variation of the patella.

A. Bipartite patella (emarginate patella, patellar subdivision, accessory patella)—a condition in which the patella appears as if a bite (notch) has been taken out of the superolateral (most frequently) corner at the insertion of the vastus lateralis muscle. Halpern and Oakley (1978) reported a possibly unique case of bilateral medial bipartite patellae. The notch is usually located at the site for attachment of a smaller accessory ossicle, the so called "patellula" in the older literature (Oetteking 1922).

Although the etiology of this condition is unknown, it is commonly

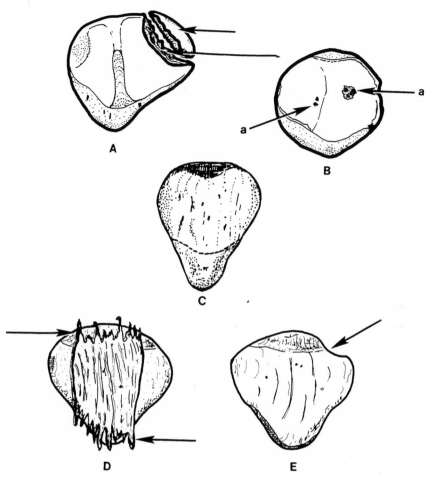

Figure 78.

believed to represent an accessory center of ossification and not the result of direct or indirect trauma to the patella (Green 1975; Halpern and Hewitt 1978; Miles 1975; Ogden 1982). Bipartite patellae occur in approximately 2 percent of the population (George 1935), are more common in males by 9 to 1 (Ogden et al. 1982), and are unilateral in 57 percent of the cases (Weaver 1977). Pain may or may not have been associated with this condition.

A review of the literature revealed that some researchers make no distinction between a bipartite patella and vastus notch (E). Some early anatomists (Kempson 1902; Todd and McCally 1921) viewed defects of the superolateral border as manifestations of a single anomalous condition (emargination) with a vastus notch representing the mildest form, and a bipartite patella the most severe form. Most researchers, however, now distinguish between a vastus notch and bipartite patella. A bipartite patella has a porous, roughened central area surrounded by smooth-bordered cortical bone for attachment of an accessory (bipartite) bone. A vastus notch, on the other hand, is a smooth surfaced, depressed or flattened area with no accompanying porosity. Bipartite patella is a common clinical, but uncommon archaeological finding. Vastus notches are very common. Adams and Leonard 1925; Callahan 1948; Giles 1928; Jones and Hedrick 1942; Miles 1975; Saunders 1978; Wright 1904.

B. Osteoarthritis of the facets—(a) the posterior surface of the patella may exhibit a raised rim at any point along its margin (OA). Other indicators of OA are surface osteophytes or surface porosity. OA of the patella is a common finding in the elderly (usually first evident in the late thirties. Todd and McCally 1921). OA in young individuals may reflect faulty/soft cartilage known as chondromalacia. Normal patellae exhibit smooth, sometimes undulating facets for articulation with the distal femur. Large, widely separated and raised oval facets may indicate paralysis and flexion contractures (bent knees) of the legs (uncommon).

C. Elongated inferior pole (dashed line denotes normal inferior patellar border)—elongation and projection of the inferior border of the patella resulting from pulling stresses at the inferior ossification center of the immature patella by the patellar tendon (Smillie 1962). Uncommon finding.

D. Ossified quadriceps tendon—exuberant bone seemingly attached to and flowing across the anterior surface of the patella. Small bony projections (enthesophytes) on the superoanterior face of the patella are common findings in most populations. However, development as shown in "D" is more likely to be seen in the elderly, typically beyond 60 years of age, and likely reflects DISH. Usually bilateral and symmetrical in appearance. If there is spiking of only the superoanterior surface of the patella the condition reflects enthesopathy, not ossification of the quadriceps tendon or osteoarthritis. Dieppe et al. 1986; Resnick and Niwayama 1988.

E. Vastus notch (emargination)—this variant appears as a notch or depression in either the superolateral (commonly) or superomedial corner of the patella. The vastus notch can be distinguished from the bipartite patella in that the surface of the vastus notch is smooth and nonporous. The size of the notch varies from nearly flat to concave. Common finding. Finnegan and Faust 1974.

FIBULA

Figure 79. Pathological conditions of the fibula.

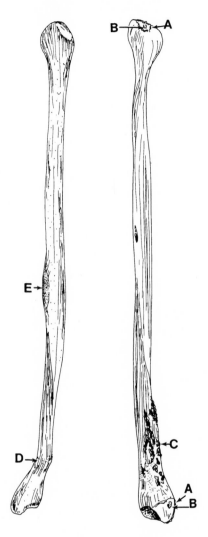

Figure 79.

A. Marginal osteophytes (OA). Common finding in the elderly.
B. Surface porosity (OA). Common finding in the elderly.
C. Exostoses due to pulling stresses of the interosseous membrane (possibly reflecting "sprains"). Common finding in middle-aged or older individuals.
D. Pott's fracture of the fibula—fracture resulting from acute trauma

from a fall or twisting of the ankle. The medial malleolus may also be fractured (not shown). Uncommon in most populations but common in individuals 60+ years (osteoporosis resulting in brittle bones).

E. Healed periostitis (x-ray would distinguish this condition from an osteoid osteoma).

Figure 80A. Bowing of the fibula due to rickets (osteomalacia, rachitis).

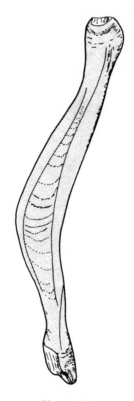

Figure 80A.

Childhood rickets typically results in lateral bowing of the long bones, especially those of the legs (upper extremity involvement is rare if the vitamin deficiency begins at an age when the children are walking). The affected bone(s) may show extreme bowing (usually most prominent in the middle third of the shaft) and flattening of the diaphysis and have expanded (flared) metaphyses. Rickets is primarily due to vitamin D deficiency during the growth period when the bones are susceptible to alteration by chemical, physiological, and mechanical factors. Osteoid fails to mineralize and bone becomes soft during remodeling associated with rapid growth. The bones become bowed when the child places weight or stress upon them in normal activities such as standing or walking (i.e., legs bow), lifting the upper body as an infant (i.e., arms bow), or simply due to the weight of the head when lying down (i.e., occipital and parietal regions may become squared or "bossed").

Other indicators of childhood rickets (see Figure 80B) are enlarged costochondral junctions of the ribs (rachitic chest, rosary-bead chest), thin cranial bones (craniotabes), barrel chest (may not be able to detect in the skeleton alone), spinal deformity including kyphosis, lordosis, or scoliosis, and a prominent sternum. (Fibulae, owing to their length and slenderness, may sometimes become bowed from soil pressure after burial.)

Vitamin D deficiency in adults (osteomalacia = softening of the bones) will result in similar but not identical bony deformities (long bone involvement rare) seen in children because the adult has ceased bone growth (not to be confused with normal bone remodeling which takes place until death). Osteomalacia can be caused by a number of factors including a diet low in calcium and phosphorus, lack of ultraviolet rays of the sun (e.g. Muslim dress codes), and intestinal malabsorption (Pitt 1981). Mankin 1974; Ortner and Putschar 1985.

Figure 80B. Skeletal involvement due to rickets.

A. Frontal bossing—flattened or "squared" frontal bones (lateral portions).
B. Parietal bossing—flattened or "squared" portions of the parietals.
C. Rachitic chest ("rosary beading")—enlarged areas (bumps) at the costochondral junction of the ribs.
D. Flared/flattened rib cage (the ribs may appear less "C" shaped than normal).
E. Flared/flattened ilia.
F. Flared metaphyses.
(Cross-hatching = frequently involved bones; stippling = infrequently involved). Note the lateral bowing of the lower limb bones.

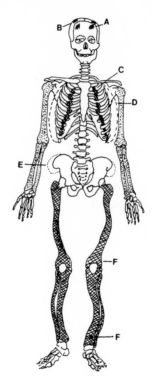

Figure 80B.

Figure 81. Fibula with chronic osteomyelitis (inset shows detail of increased vascularity [pits], striations, and periosteal new bone formation).

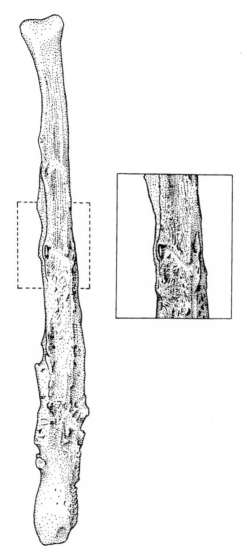

Figure 81.

FOOT

Figure 82. Squatting facets (distal tibia and talus) and an os trigonum.

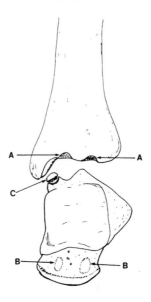

Figure 82.

A. Squatting facet—a squatting facet is scored as present whenever there is a break in the continuity of the anterior margin of the distal tibia (should be a straight line). The opposing articulating facets of the neck of the talus are represented by the hatched areas (B). Squatting facets occur when the foot is hyperdorsiflexed (toes and forefoot raised towards the knee). There may be one or two facets that are common findings in most if not all populations. Squatting facets may also be present on the talus, calcaneus, distal metatarsals, and proximal tibia. Lima et al. 1928; Ubelaker 1978.

B. Squatting facets (smooth depressions or raised tubercles) on the neck of the talus. In most cases, the areas of contact with the tibia do not show any alteration. Normal morphological trait in the human skeleton.

C. Os trigonum (nonmetric trait)—a separate small bone (accessory ossicle) of the posterior tubercle of the talus. Some researchers identify an os trigonum as such even if the bone fragment is not separate, but exhibits a fissure in the undersurface of the posterior talar facet (Steida's process). In a study of 1000 radiographs by Burman and Lapidus (1931), os trigonum was found in nearly 50 percent of the feet. A fused os trigonum will rarely fracture and become a free bone (Ihle and Cochran 1982). You must decide which criteria to use in scoring this trait as present or absent. The present authors prefer to score the trait only when a separate bone is present. McDougall 1955.

Figure 83. Congenital clubfoot (talipes or pes).

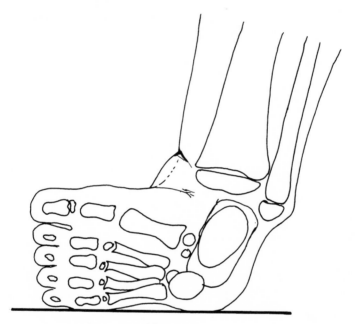

Figure 83.

Although the etiology of congenital clubfoot is unknown (Altchek 1978), the resulting foot deformity is due to malformation of the talus and medial rotation of the calcaneus. There are a number of congenital deformities of the foot including talipes equinovarus in which the foot is rotated medially and the individual walks on the "outside" of the foot. Talipes equinovarus is the most common major congenital deformity occurring once in every 1000 contemporary births (Shands 1951; Kromberg and Jenkins 1982). Other clubfoot deformities include talipes equinus (heel off the ground and walking on the toes), talipes valgus (the heel and foot are turned outward, and talipes varus (the heel is turned inward and away from the midline. Thomas 1985). The incidence of congenital clubfoot in Mongoloids is less than that

of Caucasoids (Morton 1970; Niswander et al. 1975; Wynne-Davies 1965).
Ponseti et al. 1981.

Figure 83 depicts the congenitally malformed left foot of a six-year-old
child with severe talipes equinovarus. Note that the proximal ends of the
metatarsals are overlapping and the distal ends fan out. When walking,
the individual would have placed most of the body weight on the outer
side of the fifth metatarsal (which appear thickened on x-ray). Skin
abrasions and subsequent infection often accompany untreated clubfoot
(Laura Tosi Pers. Comm. 1989). In severe cases individuals may actually
walk on the distal end (malleolus) of the tibiae.

*Figure 84. Examples of an adult malformed (A) and normal (B) right talus
(actual size).*

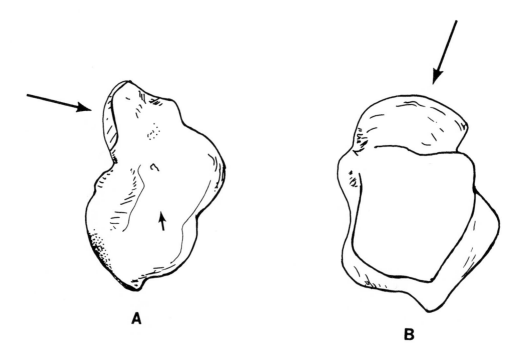

A

B

Figure 84.

Note the medial inclination of the neck and head in (A). The malformed talus is smaller than normal and the usually large, convex
posterior facet for articulation with the tibia (small arrow) is flattened.
The large arrows point to the approximate center of the articular surface

for the navicular; this surface on the malformed talus has shifted medially resulting in talipes equinovarus.

Figure 85. Pathological conditions of the calcaneus (upper drawing = lateral view and lower drawing = from above).

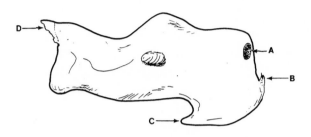

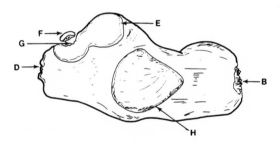

Figure 85.

A. Cyst—visible on x-ray as a radiolucent (dark) area. Rare finding possibly representing xanthoma or giant cell tumor. Dieppe et al. 1986.
B. Enthesophytes of the Achilles tendon attachment (posterior calcaneal spur)—bone spurs at this site are due to repeated or acute trauma (bleeding and inflammation) of the Achilles tendon-common finding. There also may be erosion and bony projections at this site indicative of Reiter's disease, an uncommon to rare finding (consult a radiologist). Bassiouni 1965; Dieppe et al. 1986; Hertzler 1931; Resnick and Niwayama 1988.
C. Enthesophyte (inferior calcaneal spur, heel spur)—bony projections resulting from repeated or acute trauma to the abductor hallucis and flexor digitorum brevis tendon attachments. Stresses at this site result in the lifting of the periosteum with concomitant inflammation and bone

formation. Pain is usually only associated with inferior spurs during the inflammatory stage (i.e., when the trauma occurred, Michael R. Droulette Pers. Comm. 1989). The inclination of the calcaneus is such that inferior spurs are nonweightbearing and do not come in contact with the ground. Calcaneal spurs show no sex predilection, but the frequency of occurrence increases with age. Hough and Sokoloff 1989. Common finding in the elderly.

D. Osteophytes around the margins of the cubocalcaneal joint (Lisfranc's joint). Common finding in the elderly. If there are large projecting osteophytes along this margin, they may represent ligamentous ossification.

E. Marginal osteophytes of the anterior and medial facets. Common finding in the elderly.

F. Calcaneus secundarius (Gruber 1871; Anderson 1988), secondary os calcis (Dwight 1907b)—an accessory bone (secondary center of ossification) that coalesces with the anterior calcaneal facet that, in the living, is separated from the calcaneus by a zone of cartilage. Although the actual bone fragment of a calcaneus secundarius (F) is rarely recovered in archaeological remains, the notch (G) will be visible. The surface of a calcaneal notch (G) appears as if a bite has been taken out of the anterior facet. The notch will be concave, roughened, and porous. The presence of porosity serves to distinguish a calcaneus secundarius from normal variation in the shape of the anterior calcaneal facet. Research by one of the authors (RWM) revealed that this trait is probably present in all populations and with a frequency ranging from 1.4 to 6.0 percent (Mann [in press], 1989).

G. Calcaneus secundarius notch (indicative of a calcaneus secundarius). Common finding.

H. Marginal osteophytes of the posterior facet. Common finding in the elderly.

Figure 86. Leprosy of the feet (the talus and calcaneus not shown).

Note the destruction of the metatarsals and phalanges. Many of the foot bones become "penciled" or cupped and pegged (metatarsals and phalanges), deformed, and misaligned. The nutrient foramina, especially those of the phalanges may be enlarged. Frostbite and rheumatoid arthritis, however, can result in similar deformities. The tibia and fibula may also become involved and exhibit bilateral symmetrical inflammatory changes including pitting and marked periosteal new bone formation (Manchester 1989). If this condition is encountered, seek the advice

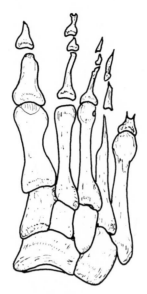

Figure 86.

of a specialist. Anderson 1982; Kulkarni and Mehta 1983; Moller-Christensen 1953, 1961; Paterson 1956.

Figure 87. Pathological conditions of the metatarsals and phalanges.

A. Periostitis of the shaft—results from a number of causes including fractures (termed "march" or "marching" fractures in the military), acute trauma with tearing of the periosteum due to a fall, or dropping an object on the foot. If the fracture is complete, there will probably be angulation (bending) of the bone. A healed fracture (a) usually appears as a bump or swollen area and will not exhibit fibrous, actively remodeling bone (callus). Common to uncommon finding.

B. Healed fracture with spicular periostitis (b)—any fracture can result in exuberant bone at the site of the break. Usually, however, the large callus associated with the early stages of healing will resorb (absorb) and reduce in size over time. However, bony spicules may remain long after the fracture has healed. Uncommon finding.

C. Bony ankylosis (c) and enlarged nutrient foramen (d)—the two conditions are unrelated. Bony ankylosis results when cartilage is destroyed, the joint space is reduced, and the opposing joint surfaces unite (fuse). This is a common finding in the distal phalanges of the feet and less so in the hand. May be age related or the result of trauma, osteoarthritis, or

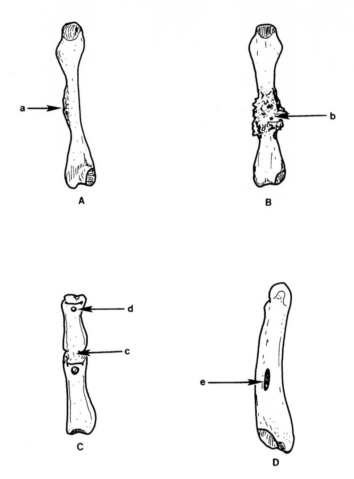

Figure 87.

other inflammatory conditions. Enlarged nutrient foramina may be encountered in the middle phalanges of the hands and feet. This condition may be the result of leprosy (rare in most archaeological samples in the U.S.), old age or other factors. Moller-Christensen 1953, 1967.

D. Enlarged nutrient foramina (e) is an uncommon finding in the metatarsals. Typically, adjoining metatarsals will be affected (may be symmetrical); for example, in one individual the left fourth and fifth metatarsals may be the only two bones affected. Fink et al. (1984) reported that normal foramina may attain a size of 1 mm. Although the etiology of enlarged foramina in the foot is unknown, such a condition in the phalanges of the hand has been found in association with thalassemia major (Poznanski 1974), Gaucher's disease (Fink et al. 1984), and leprosy

(Moller-Christensen 1967). It is possible that increased blood flow to the extremities is responsible for such enlargement.

Figure 88. Os tibiale externum (accessory ossicle). A small secondary bone attached to or articulating with the medial surface of the foot navicular. Rare to uncommon finding. Grant 1972.

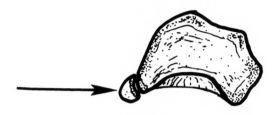

Figure 88.

Chapter VI

PSEUDOPERIOSTITIS AND CORTICAL DEFECTS IN INFANTS AND SUBADULTS (ACTUAL SIZE)

R apidly growing bones in neonates (birth to 28 days), children, and adolescents can mimic periostitis. In examining subadult bones, great care must be taken not to mistake the striations and pits associated with normal new bone growth for that of disease. In some cases this is a difficult task. In particular, the skull and long bone ends are easily incorrectly identified as diseased because of the similarity of normal growth (remodeling) with periostitis (in questionable cases seek the assistance of an experienced paleopathologist). Typically, periostitis can be differentiated from normal new bone by its "bone on bone" appearance. Again, periostitis will have a pitted or woven appearance similar to a fine sponge; normally remodeling bone along the "cut back" zone of the metaphysis may simulate abnormal remodeling. The following bones were chosen as examples of normal bone growth (remodeling) at the long bone ends (metaphyses) and the frontal bone of a neonate (newborn).

Figure 89. Normal frontal bone of a newborn (actual size).

A. Anterior fontanelle (soft spot)—normal gap in the frontal bone that closes early in childhood.
B. Normal porosity along the metopic suture.
C. Metopic suture as it closes (fuses).
D. Normal striations.
E. Normal tiny grooves.

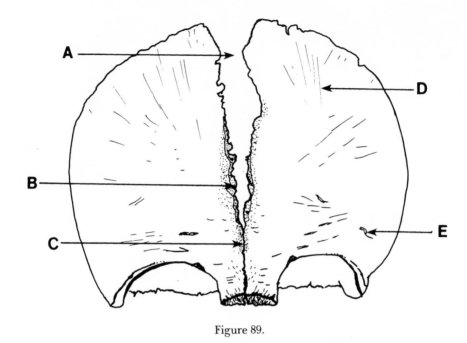

Figure 89.

Figure 90. *Normal humerus, femur, and tibia of a newborn (actual size).*

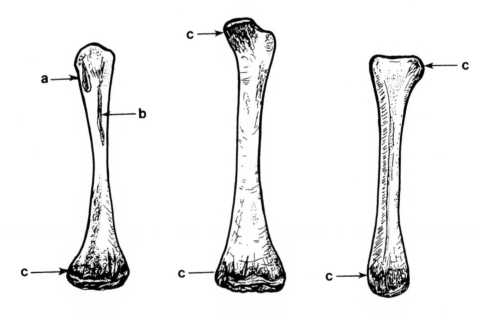

Figure 90.

a. Normal groove for attachment of the teres major muscle.
b. Normal groove for attachment of the pectoralis major muscle.
c. Normal porosity and roughened surface of rapidly growing bone.

Figure 91. Cortical defects in a subadult (14-year-old male, right proximal fibula, tibia, and radius; actual size).

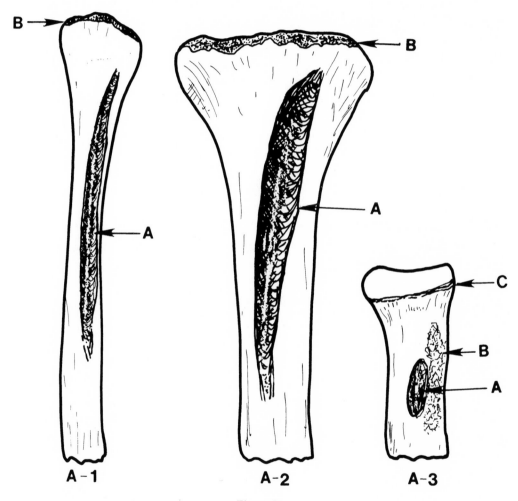

A-1   A-2   A-3

Figure 91.

Benign cortical defects (Caffey 1972; Keats 1973) appear as grooves or furrows at ligamentous or tendinous attachment sites (muscle and tendon fibers). It is likely that these grooves result from rapid normal bone remodeling and pulling stresses in the immature skeleton. Benign cortical defects represent normal variants in growing bone, regress spontaneously, and are not seen in adults (Brower 1977). However, on rare occasion the present authors have seen deep cortical defects in young but fully developed adults (for example, American soldiers killed in the Battle of Ft. Erie, War of 1812).

A recent preliminary examination by the authors of approximately 1000 subadult and adult humeri at the Smithsonian Institution revealed that in general: (1) the teres major muscle insertion is more pronounced in children up until about age four or five years, (2) during adolescence the pectoralis groove typically becomes deeper and longer than the teres groove, (3) males show a much higher frequency of both grooves than do females, and (4) the largest humeri don't necessarily have the most pronounced cortical defects. Although any muscle or tendon insertion site can be affected, the following bones were chosen as examples.

A1. Proximal fibula—cortical defect (A) for attachment of the popliteus muscle (arrow = metaphyseal plate).

A2. Proximal tibia—cortical defect (A) for attachment of the popliteus muscle (arrow = metaphyseal plate).

A3. Proximal radius—cortical defect (A) for attachment of the biceps muscle. The roughened area (B) is also a normal response to rapidly growing bone. The epiphyseal growth plate is designated by (C).

Chapter VII

FUNGAL INFECTIONS

F ungi are plant-like organisms that, for the most part, are not harmful to man. In fact, they play an important role in the decay of organic matter. Only a small group of fungi and fungi-like organisms are pathogenic (disease causing) to humans. The crucial factor is the patient's immune response.

Bone lesions in many fungal infections are similar in appearance (e.g. blastomycosis, coccidioidomycosis, and cryptococcosis). Most infections invade the bone by direct extension from a soft tissue lesion. The fungi then burrow into the bone producing a cratered or scooped-out ("icecream" scoop) lesions that appear as small (0.3 to 1.5 cm), round or ovoid depressions with thin, sharp margins. The lesions tend to be clustered and confluent with minimal periosteal reactive bone. In older lesions the bone may be exhibit marked thickening, small spicular bony projections, and shallow craters.

It is extremely difficult and often impossible to distinguish one form of mycosis from another without soft tissue and a patient's history. Differential diagnosis in mycotic infection depends largely on the patterning of skeletal involvement and fungi associated with the geographic region where the skeleton was recovered.

Figure 92. Geographic distribution of fungal infection in the continental United States (From Rippons 1980; Youmans et al. 1980).

The following are a few of the more commonly encountered fungal organisms or organisms that produce a skeletal response similar to fungi:

Actinomycosis (lumpy jaw, leptothricosis, streptotricosis)—a chronic granulomatous infection caused by the fungal-like ("higher" bacteria) Actinomyces israelii. Actinomycosis has a worldwide distribution. There are three anatomical sites of primary infection—cervicofacial (most common), thoracic, and abdominal. The skeletal area affected depends on where the organism enters the body. The lesion will almost always start as a periosteal reaction (periostitis) and new bone formation with

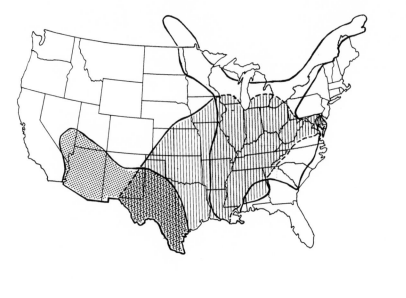

⊠ COCCIDIOIDOMYCOSIS   ▥ HISTOPLASMOSIS   ☐ BLASTOMYCOSIS

Figure 92.

Table 1.
SELECTED DISEASES THAT PRODUCE BONY REACTIONS
SIMILAR TO MYCOTIC AND ACTINOMYCOTIC INFECTION.

Bones Involved	Differential Diagnosis
Vertebrae	Tuberculosis, brucellosis, non-specific osteomyelitis, metastatic carcinoma
Long bones	Nonspecific osteomyelitis, tuberculosis, treponematosis, various neoplasms
Skull	Treponematosis, tuberculosis, multiple myeloma, metastatic carcinoma, eosinophilic granuloma, histiocytosis X, osteomyelitis

eventual erosion of the underlying cortex. Although any bone is susceptible, the mandible and vertebrae are most commonly affected. Actinomycosis of the spine has often been incorrectly identified as tuberculosis
(Simpson and McIntosh 1927). However, unlike tuberculosis, actinomycosis frequently results in erosion of the articular facets, laminae, transverse and spinous processes. Goldsand 1989; Ortner and Putschar 1985.

   Blastomycosis (North American blastomycosis, Gilchrist's disease,
Chicago disease)—a chronic granulomatous fungal infection that starts
as a pulmonary or cutaneous inflammation that often disseminates to a

systemic disease. Blastomycosis is a soil-borne fungus that grows on decaying material and is most commonly found in the southeastern and central United States (Mississippi and Ohio River basins). Males have a higher incidence than females with the average age of onset between 20 and 50 years (Jaffe 1975). Twenty to fifty percent of people with blastomycosis develop skeletal lesions. While all bones are susceptible, the most common sites are the vertebrae, skull, ribs, tibiae, tarsals, and carpals (Jones and Martin 1941). Kelley and Eisenberg 1987; Moore and Green 1982.

Coccidioidomycosis (coccidiomycosis, Posadas disease, valley fever, desert rheumatism, San Joaquin fever)—caused by the inhalation of soil-borne fungal spores, this disease is generally a benign inflammation of the respiratory tract. Only a small number of people (usually immune deficient) will contract the disease (Fiese 1958). Coccidioidomycosis is endemic to a well-defined area of the southeastern United States (California, New Mexico, Arizona, and southwestern Texas), Mexico, Central and South America (Binford and Conner 1976). All bones are potential sites with the ribs, vertebrae, skull, and bones of the extremities favored (Fiese 1958). Periosteal new bone formation and sequestra are uncommon. The lesions tend to be lytic and favor the cancellous areas of bone. Bried and Galgiani 1986; Hoeprich 1989.

Histoplasmosis (Darling's disease, cave disease, Ohio Valley disease, Tingo Maria fever)—a common pulmonary infection with worldwide distribution (Binford and Conner 1976). The fungus grows in soil (particularly soil rich in bird feces) and enters the body through the inhalation of spores. Histoplasmosis is endemic in the eastern and central United States. Youmans et al. 1980. For further information refer to Attah and Cerruti 1979; Brook et al. 1977; Caravio et al. 1977; Dalinka et al. 1971; Daveny and Ross 1969; Echols et al. 1979; Greer 1962; Poswall 1976; Procknow and Loosli 1958; Reeves and Pederson 1954; Seligsohn et al. 1977; Wheat 1989.

Chapter VIII

TREPONEMATOSIS

Treponematosis is a chronic infection caused by a corkscrew-shaped organism (spirochete) of the genus Treponema (meaning "twisted thread") and in Europe and the United States is commonly known as syphilis (lues or leutic disease). Due to the clinical and geographical variation the infection is divided into four types—pinta, yaws, endemic syphilis, and venereal (and congenital) syphilis (Table 2). If treponematosis is suspected, seek the advice of someone familiar with the disease. Diagnosing bone as syphilitic (or the other forms) is difficult due to the similarity of these conditions with nonspecific osteomyelitis and other allied diseases (syphilis is often called the "great imitator". Thomas 1985). The following information is offered only as an introduction to the treponematoses and is not meant to serve as a tool for diagnosis in atypical cases.

Pinta (Treponema carateum) is the least destructive and most benign form of treponematosis. Pinta (Spanish for "spot," "dot" or "mark") is endemic only to the American Tropics and is most frequently found in Mexico and parts of Central and South America. Pinta is spread through direct nonsexual contact that initially appears as a skin rash, undergoes pigment changes, and spreads (secondary pinta). The tertiary (final) stage is usually marked by a general depigmentation of the lesions. Pinta is the only treponemal infection that does not cause bone lesions.

Yaws (Treponema pertenue), like pinta, is a nonvenereal endemic juvenile disease (Hudson 1958a) that is spread through close contact with an infected sore (e.g. children playing). Yaws is found worldwide in hot and humid tropical regions, usually begins in childhood, and starts as a painful localized lesion/sore on an exposed extremity (Binford and Connor 1976). If the sore is located near bone it will induce periostitis (polydactylitis is a frequent finding in early stage yaws. Hoeprich 1989). Bone lesions in yaws are rare (ca. 15%) and generally considered indistinguishable from other treponemal infections (Hackett 1976). One of the most consistent findings in yaws is anterior bone hypertrophy of the tibia ("boomerang leg"); only rarely does the fibula show this feature. In the late stages, gummatous periostitis and osteomyelitis develop which are almost identical (although not as severe) to tertiary syphilis. Synonyms are bouba, frambesia, tropica, parangi, and pian (Thomas 1985).

Endemic syphilis (treponarid, bejel) is a nonvenereal infection found in warm, arid to semiarid environments (e.g. Africa, the Middle East, and Asia). Endemic syphilis is spread through close contact with infected persons (e.g. kissing) or contaminated objects (e.g. drinking utensils). Binford and Connor (1976) propose that the clinical symptoms of endemic syphilis and yaws are so similar that the infectious agent should be classified as Treponema pertenue (yaws) instead of T. pallidum (venereal and endemic syphilis). Steinbock (1976) suggests that the endemic form is an intermediate manifestation between yaws and venereal syphilis. Hudson (1958b) writes " . . . the standard of living and the level of hygiene in a given community determines whether its syphilis will be venereal or non-venereal." Regardless, bone lesions are uncommon and are difficult to distinguish from yaws and venereal syphilis. Current research by Donald J. Ortner suggests the development of joint lesions in some cases of yaws that are not seen in syphilis.

Venereal syphilis (Treponema pallidum, "pale twisted thread"), as the name implies, is transmitted through sexual contact. Venereal syphilis is found worldwide and is the only treponemal infection that can be passed from mother to fetus (congenital syphilis). The venereal form, also known as acquired syphilis, uncommonly involves bone (10 to 20%), usually in the tertiary stage of the disease, and two to ten years after the initial infection (Ortner and Putschar 1985; Steinbock 1976). The bones most commonly affected are the tibia and skull although any bone(s) may be involved (Jaffe 1975).

In the skull, syphilis almost always first attacks the outer table, unlike tuberculosis and some neoplasms that start in the diploe and work outwards. The most characteristic cranial lesion is the pattern of scarring seen on the frontal and parietals (occasionally affects the occipital and temporals), called caries sicca (pronounced "sick-uh"). There are three types of lesions that together form caries sicca: stellate scarring, nodes, and cavitations (Hackett 1976). Cavitations appear as areas of depressed bone that usually do not perforate the inner table. The walls and rims of the depressions are comprised of smooth sclerotic bone (nodes) and stellate (star-like) scars, also known as radial scars. These scars appear as smooth-rimmed furrows that radiate from a central point. Caries sicca represents a healed stage of the disease and will remain visible through-out life. Active areas of early inflammation may be difficult to differenti-ate from other infections such as tuberculosis, osteomyelitis, neoplasm, and others.

Congenital or prenatal syphilis (Rudolph 1989) results from infection of the fetus by the mother during intrauterine development (after the 16th intrauterine week). The infection often kills the child before birth while milder cases (syphilis congenita tarda) may remain dormant for many years. Spread of the spirochete usually results in massive bone infection and enlargement. Bone inflammation (panosteitis) is frequently detectable radiographically at the distal radius and ulna (Krugman et al. 1985). The following osseous symptoms (stigmata) were noted upon examination of 271 native-born Americans at the Boston City Hospital (Fiumara and Lessell 1970) and 100 patients on the islands in the Carib-bean (Fiumara and Lessell 1983) who were diagnosed with late congeni-tal syphilis (Table 2):

Table 2.
OSSEOUS CHANGES PRESENT IN PATIENTS
WITH LATE CONGENITAL SYPHILIS
(Fiumara and Lessell 1970, 1983).

Stigmata	*Boston (1960–1969) n/%*	*Caribbean (1975–1981) n/%*
Hutchinson teeth	171/63.1*	54/100**
Mulberry molars	176/64.9*	71/100**
Frontal bossae	235/86.7	96/96
Short maxillae	227/83.3	100/100
Sternoclavicular thickening	107/39.4	81/81
Relative protuberance of the mandible ("Bull-dog-Jaw")	70/25.8	84/84
High palatal arch	207/76.4	79/79
Saddle nose	199/73.4	92/92
Sabre shin	11/4.1	9/9
Scaphoid scapulae	2/0.7	3/3

*Not a true picture of the incidence because many patients had had their teeth extracted or were edentulous.
**The remainder were edentulous.

For further information refer to Baker and Armelagos 1988 (historical overview); Hackett 1976; Hudson 1958a, 1958b; Murray et al. 1956; Ortner and Putschar 1985; Stewart and Spoehr 1952; Steinbock 1976; Suzuki 1984.

Chapter IX

TUMORS

A tumor, or neoplasm, is new growth of tissue that is uncoordinated with the normal tissue and may interfer with normal physiology (Tables 3–4). Tumors may be either benign or malignant. Benign tumors are usually slow growing, localized, noninvasive, and usually present no serious threat to the life (e.g. osteoma). Malignant tumors, on the other hand, grow rapidly, spread (metastasize) to other tissue, and often eventually result in death of the individual (e.g. osteosarcoma). Invasion of other tissue by metastasis can occur either through the lymphatic system, blood stream, or by direct extension of the tumor. Not all tumors fall clearly into these two categories. Giant cell tumors, for example, are generally considered benign but up to 30% develop malignant tendencies (Hutter et al. 1962).

Tumors are classified according to the tissue of origin or differentiation such as bone or cartilage. The prefix usually indicates the tissue of origin (e.g. osteo = bone, chondro = cartilage). However, not all tumors that affect bone arise from bone and cartilage. For example, liposarcoma (from fat), leiomysarcoma (smooth muscle), fibroma (fibrous tissue), hemangioma (vascular tissue), and neurofibroma (nerve tissue) to name a few.

It should be remembered that differentiating one type of tumor from another disease condition is extremely difficult. This task is made even more difficult when attempting a diagnosis in dry bone without benefit of patient history, soft tissues, or the entire skeleton. For example, was there pain or soft-tissue swelling associated with the tumor? Did the individual sustain trauma at the tumor site? Which bones of the skeleton were affected? All of these considerations must be taken into account if a correct interpretation is to be made. Even with a well documented patient history an accurate diagnosis is often only narrowed down to a few possibilities that best fit the clinical "picture." If a neoplasm is suspected, first radiograph the bone(s), refer to Tables 3 and 4 and Figure

93, and seek the assistance of a qualified paleopathologist, orthopaedic radiologist or orthopaedic pathologist for the final interpretation.

Sources: Brothwell and Sandison 1967; Coley 1949; Dahlin 1967; Grupe 1988; Hutter 1962; Jaffe 1958; Manchester 1983; Suzuki 1987; Tkocz and Bierring 1984.

Figure 93. Composite diagram illustrating frequent sites of bone tumors. The diagram depicts the end of a long bone which has been divided into the epiphysis, metaphysis, and diaphysis. The typical sites of common primary bone tumors are labeled.[*]

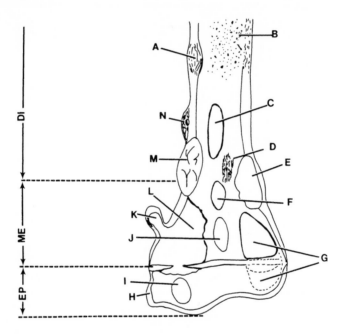

Figure 93.

A. Cortical fibrous dysplasia, adamantinoma.
B. Round cell lesions—Ewing's Sarcoma, reticulum cell sarcoma, myeloma.
C. Fibrous dysplasia.
D. Fibrosarcoma.
E. Fibroxanthoma (fibrous cortical defect: non-ossifying fibroma).

[*]Adapted from Madewell, J.E. et al.: Radiologic and Pathologic Analysis of Solitary Bone Lesions: Part I: Internal Margins. *The Radiologic Clinics of North America* 19(4), W. B. Saunders Company, Philadelphia, Pennsylvania, 1981.

Table 3.
TUMOR AND TUMOR-LIKE CONDITIONS OF BONE BY REGION.

	Benign	Malignant	Blastic	Lytic	Location
LONG BONES					
Adamantinoma		X		X	D*
Aneurysmal Bone Cyst	X			X	M
Chondroblastoma	X			X	E/M
Chondroma (enchondroma)	X			X	D/M
Chondrosarcoma		X	X	X	M
Eosinophilic Granuloma	X		X	X	M/D
Ewing's Sarcoma		X	X	X	D/M
Fibrosarcoma		X		X	M
Fibrous Dysplasia	X		X	X	D/M
Fibrous Cortical Defect	X			X	M
Giant Cell Tumor	X			X	E/M
Metastatic Carcinoma		X	X	X	M/D
Multiple Myeloma		X		X	M/D
Non-ossifying Fibroma	X			X	M
Osteoid Osteoma	X		X	X	D
Osteoblastoma	X			X	D/M
Osteochondroma	X		X		M
Osteosarcoma		X	X	X	M
Solitary Bone Cyst	X			X	M
RIB/STERNUM					
Aneurysmal Bone Cyst	X			X	Both
Chondroblastoma	X			X	Ribs
Chondrosarcoma		X	X	X	Both
Fibrous Dysplasia	X		X	X	Ribs
Metastatic Carcinoma		X	X	X	Both
Multiple Myeloma		X		X	Both
Osteoid Osteoma	X		X	X	Ribs
Osteosarcoma		X	X	X	Both
PELVIC/PECTORAL					
Aneurysmal Bone Cyst	X			X	All
Chondroblastoma	X			X	All
Chondrosarcoma		X	X	X	P/S
Metastatic Carcinoma		X	X	X	All
Multiple Myeloma		X		X	P
Osteoid Osteoma	X		X	X	P/S
Osteosarcoma		X	X	X	All

Table 3. Continued

	Benign	Malignant	Blastic	Lytic	Location
SPINE					
Aneurysmal Bone Cyst	X			X	All
Hemangioma	X			X	T/L
Metastatic Carcinoma		X	X	X	All
Multiple Myeloma		X		X	All
Osteoblastoma	X			X	All
SKULL					
Eosinophilic Granuloma	X		X	X	Calv.
Fibrous Dysplasia	X		X	X	All
Hemangioma	X			X	Calv.
Metastatic Carcinoma		X	X	X	All
Multiple Myeloma		X		X	Calv.
Osteoblastoma	X			X	All
Osteoma	X		X		Calv.
HAND/FOOT					
Chondroblastoma	X			x	Tars.
Chondroma (enchondroma)	X			X	D
Metastatic Carcinoma		X	X	X	All
Osteoblastoma	X			X	All
Osteochondroma	X		X		All
Osteoid Osteoma	X		X	X	All

*Diaphysis; E = epiphysis; M = metaphysis; P = pelvis; S = scapula; T = thoracic; L = lumbar.

F. Bone cyst, osteoblastoma.
G. Giant cell tumor—child: metaphyseal
                     adult: "end of bone".
H. Articular osteochondroma (dysplasia epiphysealis hemimelica).
I. Chondroblastoma.
J. Enchondroma, chondrosarcoma.
K. Osteochondroma.
L. Osteosarcoma.
M. Chondromyxoid fibroma.
N. Osteoid osteoma.
DI = diaphysis
ME = metaphysis
EP = epiphysis

Table 4.
TYPICAL AGE RANGE OF THE OCCURANCE OF BONE TUMORS

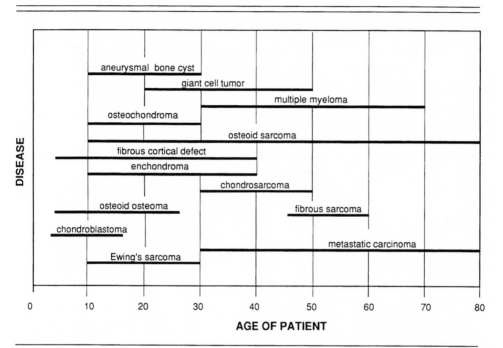

(Reprinted with permission from Griffiths, *Basic Bone Radiology,* Appleton-Century-Crofts, Norwalk, CT, 1981, as modified).

Chapter X

HYDATID CYST

Figure 94. Hydatid cyst (hydatid disease. Rong and Zhang 1985)—echino-coccus cyst; hydatidosis (actual size).

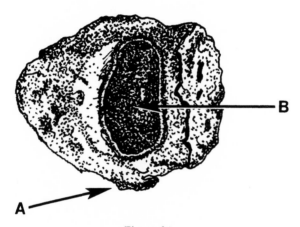

Figure 94.

The hydatid cyst represents the larval stage in the evolution of the tape worm, usually Echinococcus granulosus (Thomas 1985). Human hydatid disease caused by E. granulosus is an important zoonosis (disease communicable from animals to man) in many areas of the world (Matossian et al. 1977). In the wild, herbivores such as caribou, moose, antelope, and deer may harbor the parasite as an intermediate host with man and wild carnivores (e.g. wolves, jackals, and lions) as final hosts. In the host, the cyst is white, spherical and filled with fluid. Adult worms are four segmented, measure 3–5 mm in length, are only seen in dogs and closely related animals (e.g. wolves), and cannot develop in man (Markell and Voge 1981).

Hydatid cysts grow at a rate of 1 mm per month (Noble and Noble 1982), may reach 6 inches (15 cm) in diameter, and usually present no symptoms until it is at least 10 cm in diameter. Hydatids may have walls an inch or more thick (Ray 1860). The outer surface of the cyst calcifies

153

(Schmidt and Roberts 1977), is irregular (A), and may have an opening (B) where it attached to the soft tissue within an organ (likely the liver, lung, muscles, spleen, and even bone [2%]. Sparks et al. 1976). Any number of cysts may be found within the chest or abdominal cavities of an individual.

Hydatid cysts (in humans) have a worldwide distribution but are more common in many groups from the Mediterranean, East Africa, USSR, the Middle East with an incidence as high as 20 percent in parts of South America where sheep herding predominates. Squire 1964; Szypryt et al. 1987; Martin Wolfe Pers. Comm. 1988.

Chapter XI

PERI-MORTEM VERSUS
POST-MORTEM FRACTURES

Figure 95. Peri-mortem and post-mortem fractures (tibia used as example).

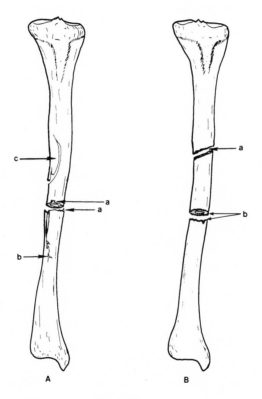

Figure 95.

Forensic pathologists refer to soft tissue injuries usually as ante-mortem (inflicted during life) or post-mortem (occurring after death). The degree of tissue reaction and bleeding are the primary indicators. Occasionally wounds fall in between, inflicted while the individual was in the act of dying, and this is termed peri-mortem. For soft tissues, this period is rather short and precise; in dry bones this term has a slightly different

155

meaning and covers a time which is indistinguishable from ante-mortem and may persist until weeks after death. The reason for this is two-fold. First, many times there is no soft tissue left to aid ante-, peri-, or post-mortem determination. Second, living bone is a visco-elastic material with special fracture properties that do not disappear until the elastic (collagen) component deteriorates after death and this may take weeks or months. Thus, in the absence of soft tissue evidence, the peri-mortem period loses its precision and must include recent ante-mortem events and an obscure post-mortem period during which the bone retains its "fresh" visco-elastic properties.

A. Ante-mortem/Peri-mortem fractures generally leave sharp, smooth, often beveled fracture lines (a). Most of the bone at the site of fracture will be present. Fractures of fresh bone often have associated radiating fracture lines at the site of trauma (b). The fracture ends are usually as discolored as the adjacent surface bone (old, dirty appearing). Displaced, curved but still adhering bone fragments or splinters or incomplete fractures with bending of the bone (greenstick fractures) (c), and dirt within the breaks are also associated with peri-mortem trauma. But it is not always possible to distinguish peri-mortem fractures from post-mortem. Further, because the breakage may have occurred weeks to months after the person's death, but while the bone was still fresh, statements establishing relationship to the cause of death (e.g. gunshot) must be made with caution and judgement. As documented by the authors' research on Civil War soldiers, fracture repair may take two weeks before it is apparent in dry bone. So, regarding dry bones, peri-mortem may extend from two weeks before death to months after death.

B. Post-mortem fractures usually refer to events that occur after death, and clearly indicates that the bone has lost its elastic properties. Varied conditions of decomposition impact the timing of the loss of bony elasticity. When bone is exposed to the environment for a long period of time it becomes dry, weathered, subject to distortion, breaking, checking, and cracking. Although it is extremely difficult to ascertain what may have caused a particular break, it is often possible to discern post-mortem features. Post-mortem characteristics reflect its viscous or brittle nature, fracture edges are irregular to jagged but with blunt/dull edges (a), little beveling, few or no tiny radiating fractures, and small areas of missing bone that became "dust" upon breakage (b).

Figure 96. Healed fracture (complete) of the left fibula.

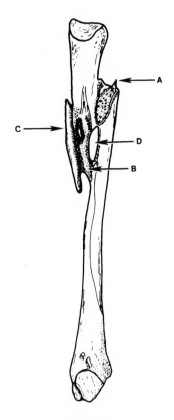

Figure 96.

Although the fibula was chosen as an example of a healed displaced long bone fracture, the fracture terminology applies to any long bone in the body. In this example, the fracture occurred at the proximal third of the shaft and resulted in shortening and overriding of the two broken ends (side-to-side displacement). Note the displaced bone fragment (C) and opening between the fused ends of the broken shaft (D). Although the fracture is nearly healed, the pointed portion of the shaft (A) suggests that further resorption may occur. If this happens, the point will become dull, rounded, and the cortex at the fracture site may eventually smooth over. Evidence of healing can be seen by fusion of the proximal third of the shaft to the middle third of the shaft (B). Extensive callus formation may accompany a fracture. If the bone appears "swollen" and radiographs reveal large areas of sclerosis, it is possible that the fracture site was infected (osteomyelitis) (Resnick and Niwayama 1981). Rockwood and Green 1975; Rogers 1982.

Although there is disagreement as to how long it takes for the earliest signs of periosteal remodeling (callus formation) to become visible on the surface of the cortex, it is probably safe to say that any active remodeling visible in dry bone suggests at least ten days of healing. Fractures are uncommon to common finding.

Figure 97. Types of long bone fractures.

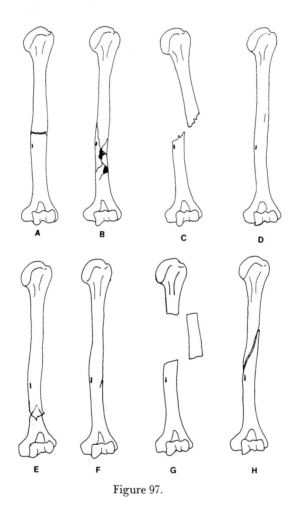

Figure 97.

A. Transverse—fracture in which the bone is broken perpendicular to its long axis (see Greenstick fracture). If the bone is broken in two (A) the fracture is termed complete.

B. Comminuted—fracture in which the bone is broken into many pieces or fragments. Common finding in the elderly.

C. Oblique and displaced—an oblique fracture is one in which the bone is completely broken at an angle diagonal to its long axis. Displaced fractures may result in misalignment and overriding of the broken ends. Improper or inadequate immobilization of the fracture may result in a pseudarthrosis or loss in bone length.

D. Hairline—minor fracture in which the bone fragments remain in perfect alignment (e.g. "march").

E. Impacted—fracturing and subsequent wedging of one bone end into the interior of another. Impaction results in a reduction in the normal length of the bone.

F. Incomplete—fracture more severe than a hairline but less severe than a complete with no separation of bone fragments (e.g. sword wound).

G. Segmental—fracture in which a significant portion (intact segment) of the bone is displaced. Again, loss of bone or improper immobilization may result in a pseudarthrosis or loss of bone length.

H. Spiral—oblique fracture. Common finding in the elderly and usually resulting from osteoporosis and brittle bones. Spiral fractures are commonly associated with perimortem or "fresh" bone.

Stellate (not shown)—fracture in which lines radiate from the central point of impact (e.g. penetrating wound such as a gunshot).

Stress (not shown)—fine hairline fracture that may go unnoticed for weeks in the living (a frequent incidental radiographic finding). Stress fractures result from too rapid an increase in a particular activity. For example, "recruit's" fractures of the tibia result when young individuals are made to run for extended periods without properly "working up" to the activity. Common sites are the metatarsals and calcanei ("march") and tibiae ("recruit's").

Undisplaced (not shown)—fracture in which the bone fragment(s) and host bone remain in approximate anatomical position.

Greenstick (not shown)—the thick periosteum surrounding subadult long bones may result in the bending (actually, a slight fracture will be histologically visible) of one side of the cortex and fracture of the other. The appearance is similar to a broken "green stick."

Chapter XII

MUSCLE ATTACHMENTS

Figure 98. Selected muscle, tendon, and ligament attachments in the human skeleton (ventral view).

1. Temporalis
2. Masseter
3. Masseter
4. Conoid ligament
5. Trapezoid ligament
6. Pectoralis minor
7. Coracobrachialis and short head of biceps
8. Inferior belly of omohyoid
9. Long head of triceps
10. Subscapularis
11. Serratus anterior
12. Deltoid
13. Brachialis
14. Brachioradialis
15. Extensor carpi radialis longus
16. Pronator teres
17. Inguinal ligament
18. Sartorius
19. Rectus femoris
20. Psoas minor
21. Piriformis
22. Gluteus minimus
23. Vastus lateralis
24. Iliofemoral ligament
25. Vastus intermedius
26. Fibular collateral ligament
27. Patellar ligament
28. Sartorius
29. Tibialis anterior

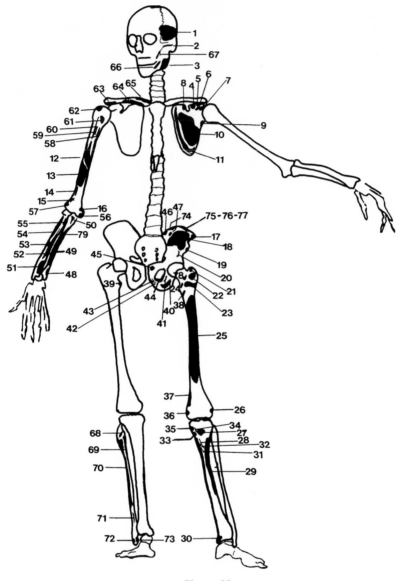

Figure 98.

30. Medial collateral ligament
31. Semitendinous
32. Gracilis
33. Tibial collateral ligament
34. Vastus medialis
35. Semimembranosus

36. Tibial collateral ligament
37. Adductor magnus
38. Vastus medialis
39. Psoas and iliacus
40. Adductor magnus
41. Quadratus femoris
42. Gracilis
43. Adductor brevis
44. Obturator externus
45. Adductor longus
46. Erector spinae
47. Iliolumbar ligament
48. Pronator quadratus
49. Flexor digitorum profundus
50. Brachialis
51. Pronator quadratus
52. Flexor pollicis longus
53. Pronator teres
54. Supinator
55. Biceps
56. Common flexor
57. Common extensor
58. Teres major
59. Latissimus dorsi
60. Pectoralis major
61. Subscapularis
62. Supraspinatus
63. Deltoid
64. Pectoralis major
65. Trapezius
66. Buccinator
67. Temporalis
68. Peroneus longus
69. Extensor digitorum longus
70. Peroneus longus
71. Peroneus tertius
72. Calcaneofibular ligament
73. Anterior talofibular ligament
74. Quadratus lumborum

75. Transversus abdominus
76. Internal oblique
77. External oblique
78. Iliofemoral ligament
79. Flexor digitorum superficialis, radial head
80. Iliacus

Figure 99. Selected muscle, tendon, and ligament attachments in the human skeleton (dorsal view).

1. Trapezius
2. Semispinalis capitis
3. Sternocleidomastoid
4. Supraspinatus
5. Infraspinatus
6. Teres minor
7. Lateral head of triceps
8. Medial head of triceps
9. Medial head of triceps
10. Brachioradialis
11. Extensor carpi radialis longus
12. Common extensor
13. Common flexor
14. Gluteus minimus
15. Gluteus maximus
16. Piriformis
17. Gluteus minimus
18. Adductor magnus
19. Medial head of gastrocnemius
20. Adductor magnus
21. Tibial collateral ligament
22. Semimembranosus
23. Popliteus
24. Medial collateral ligament
25. Posterior tibiofibular ligament/inferior transverse ligament
26. Interosseous ligament
27. Flexor digitorum longus
28. Tibialis posterior
29. Soleus
30. Fibular collateral ligament

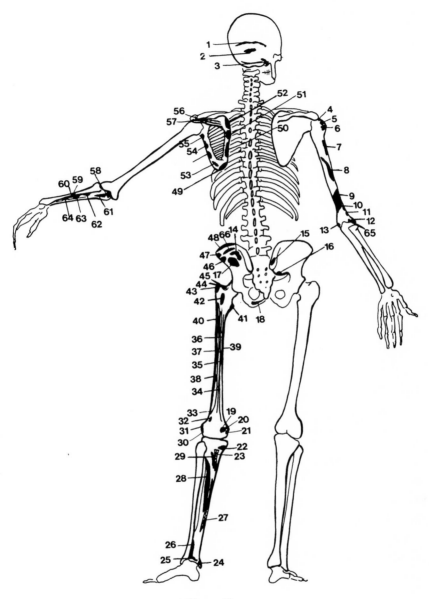

Figure 99.

31. Lateral head of gastrocnemius
32. Plantaris
33. Short head of biceps
34. Adductor magnus
35. Vastus lateralis

36. Adductor brevis
37. Vastus intermedius
38. Adductor longus
39. Vastus medialis
40. Gluteus maximus
41. Psoas and iliacus
42. Quadratus femoris
43. Gluteus medius
44. Obturator externus
45. Obturator internus
46. Sartorius
47. Inguinal ligament
48. External oblique
49. Latissimus dorsi
50. Rhomboid major
51. Rhomboid minor
52. Levator scapulae
53. Teres major
54. Teres minor
55. Long head of triceps
56. Trapezius
57. Deltoid
58. Triceps
59. Pronator teres
60. Extensor pollicis longus
61. Anconeus
62. Adductor pollicis longus
63. Apponeurotic attachment of flex. digi. prof., flex. carp. ul. and ext. carp. ul.
64. Extensor indicis
65. Anconeus
66. Tensor fasciae latae

BIBLIOGRAPHY

Adams, J. L.: The supracondyloid variation in the human embryo. *Anatomical Record 59:*315, 1934.

Akrawi, F.: Is bejel syphilis? *British Journal of Venereal Disease 26:*115, 1949.

Albright, F., Smith, P. A., and Richardson, A. M.: Postmenopausal osteoporosis: its clinical features. *Journal of American Medical Association 116:*2465, 1941.

Allbrook, D. B.: Characteristics of the East African vertebral column. *Journal of Anatomy 88:*559, 1954.

Allen, H.: *A System of Human Anatomy, Sec. II, Bones and Joints.* Philadelphia, 1882.

Altchek, M.: Congenital clubfeet (letters to the editor). *Clinical Orthopaedics and Related Research 130:*303–305, 1978.

Anderson, J. E.: The people of Fairty. *National Museum of Canada Bulletin 193:*28, 1963.

——: The osteological diagnosis of leprosy. Proceedings of the Paleopathology Association 4th European Meeting, Middleberg, Antwerpen, 1982.

Anderson, T.: Calcaneus secundarius: an osteo-archaeological note. *American Journal of Physical Anthropology 77:*529, 1988.

Angel, J. L.: Osteoporosis: thalassemia? *American Journal of Physical Anthropology 22:*369, 1964.

——: Porotic hyperostosis, anemias, malarias and marshes in prehistoric eastern Mediterranean. *Science 153:*760, 1966.

——: Colonial to modern skeletal change in the USA. *American Journal of Physical Anthropology 45:*723, 1976.

——: History and development of paleopathology. *American Journal of Physical Anthropology 56:*509, 1981.

Angel, J. L., Kelley, J. O., Parrington, M., and Pinter, S.: Life stresses of the free black community as represented by First African Baptist Church, Philadelphia. *American Journal of Physical Anthropology 74:*213, 1987.

Attah, C. A., and Cerruti, M. M.: Aspergillus osteomyelitis of the sternum after cardiac surgery. *New York State Journal of Medicine 79:*1420, 1979.

Baker, B. J., and Armelagos, G. J.: The origin and antiquity of syphilis. *Current Anthropology 29*(5):703, 1988.

Barnard, L. B., and McCoy, S. M.: The supracondyloid process of the humerus. *Journal of Bone and Joint Surgery 28:*845, 1946.

Barrie, H. J.: Osteochondritis dissecans 1887–1987: a centennial look at Konig's memorable phrase. *Journal of Bone and Joint Surgery 69*B(5):693, 1987.

Barry, H. C.: *Paget's Disease of Bone.* London, E&S Livingstone Ltd., 1969.

Bass, W. M.: *Human Osteology: A Laboratory and Field Manual,* 3rd ed. Columbia, Missouri Archaeological Society, 1987.

Bassiouni, M.: Incidence of calcaneal spurs in osteoarthrosis and rheumatoid arthritis, and in control patients. *Annals of the Rheumatic Diseases, 24:*490, 1965.

Berry, A. C., and Berry, R. J.: Epigenetic variation in the human cranium. *Journal of Anatomy 101*(2):394, 1967.

Bertaux, T. A.: *L'humerus et le femur.* Lile, France, 1891.

Binford, C. H., and Conner, D. H.: *Pathology of Tropical and Extraordinary Diseases.* Washington, D.C., Armed Forces Institute of Pathology, 1976, vol. 1.

Blackburne, J. S., and Velikas, E. P.: Spondylolisthesis in children and adolescents. *Journal of Bone and Joint Surgery 59*B:490, 1977.

Blackwood, H. J. J.: Arthritis of the mandibular joint. *British Dental Journal 115*(8):317, 1963.

Bloch, I.: *History of Syphilis, System of Syphilis.* London, Hodder & Stoughton, 1908.

Bluestone, C. D., and Klein, J. O.: *Otitis Media in Infants and Children.* Philadelphia, Saunders, 1988.

Bowen, J. R., Kumar, V. P., Joyce, John J. III, and Bowen, J. C.: Osteochondritis dissecans following Perthes disease. In Burwell, R. B. and Harrison, M. H. M. (Eds.): *Clinical Orthopaedics and Related Research 209:*49, 1986.

Bradford, D. S.: Spondylolysis and spondylolisthesis. In Chou, S. N. and Seljeskog, E. L. (Eds.): *Spinal Deformities and Neurological Dysfunction.* New York, Ravon Press, 1978.

Bradley, J., and Dandy, D. J.: Osteochondritis dissecans and other lesions of the femoral condyles. *Journal of Bone and Joint Surgery 71*B(3):518, 1989.

Bridges, P. S.: Spondylolysis and its relationship to degenerative joint disease in the prehistoric Southeastern United States. *American Journal of Physical Anthropology 79:*321, 1989.

Bried, M. J., and Galgiani, J. N.: Coccidiodides immitis infections in bones and joints. *Clinical Orthopaedics and Related Research 211:*235, 1986.

Brook, C. J., Ravikrishnan, K. P., and Weg, J. G.: Pulmonary and articular sporotricosis. *American Review of Respiratory Disorders 116:*141, 1977.

Brothwell, D. R.: *Digging Up Bones: The Excavation Treatment and Study of Human Skeletal Remains.* Ithaca, Cornell University Press, 1981.

——: The real history of syphilis. *Science Journal,* pp. 27–33, 1970.

Brothwell, D. R.,orders 116:141, 1977.

Brothwell, D. R.in Antiquity. A Survey of the Diseases Injuries and Surgery of Early Populations. Sprihaca, Cornell University Press, 1981.

——: The real history of syphilis. *Science Journal,* pp. 27–33, 1970.

Brothwell, D. R., and Sandison, A. T. (Eds.): *Diseases in Antiquity. A Survey of the Diseases Injuries and Surgery of Early Populations.* Springfield, Thomas, 1967.

Brower, A. C.: Cortical defect of the humerus at the insertion of the pectoralis major. *American Journal of Roentgenology 128:*677, 1977.

Brower, A. C., and Allman, R. M.: The neuropathic joint: a neurovascular bone disorder. *Radiologic Clinics of North America 19*(4):571, 1981.

Bulos, S.: Herniated intervertebral lumbar disc in the teenager. *Journal of Bone and Joint Surgery* 55B:273, 1973.

Bunnell, W. P.: Back pain in children. *Orthopedic Clinics of North America* 13(3):587, 1982.

Burke, M. J., Fear, E. C., and Wright, V.: Bone and joint changes in pneumatic drillers. *Annals Rheumatic Diseases* 36:276, 1977.

Burman, M. S., and Lapidus, P. W.: Functional disturbances caused by inconstant bones and sesamoids of the foot. *Archives Surgery* 22:936, 1931.

Busch, M. T., and Morrissy, R. T.: Slipped capital femoral epiphysis. *Orthopaedic Clinics of North America* 18:637, 1987.

Bushan, B., Watal, G., Ahmed, A., Saxena, R., Goswami, K., Pathania, A. G.: Giant ivory osteoma of frontal sinus. *Australas Radiology* 31(3):306, 1987.

Caffey, J.: *Pediatric X-ray Diagnosis,* 6th ed. Chicago, Year Book Medical Publishers, 1972.

Calin, A.: Ankylosing spondylitis. *Clinics in Rheumatic Diseases* 11(1):41, 1985.

Campillo, D.: Herniated intervertebral lumbar discs in an individual from the Roman Era, exhumed from the "Quinta de San Rafael" (Tarragona, Spain). *Journal of Paleopathology* 2(2):89–94, 1989.

Caraveo, J., Trowbridge, A. A., Amaral, B. W., Green, J. B., Cain, P. T., and Hurley, D. L.: Bone marrow necrosis associated with a mucor infection. *American Journal of Medicine* 62:404, 1977.

Carlson, D., Armelagos, G., and Van Gerven, D.: Factors influencing the etiology of cribra orbitalia in prehistoric Nubia. *Journal of Evolution* 3:405, 1974.

Carney, C. N., and Wilson, F. C.: Infections of bones and joints. In Wilson, F. C. (Ed.): *The Musculoskeletal System.* Philadelphia, Lippincott, 1975.

Casscells, S. W.: The arthroscope in the diagnosis of disorders of the patellofemoral joint. *Clinical Orthopaedics and Related Research* 144:45, 1979.

Caterall, A.: *Legg-Calve-Perthes Disease.* New York, Churchill Livingstone, 1982.

Cave, A. J. E.: The nature and morphology of the costoclavicular ligament. *Journal of Anatomy* 95:170, 1961.

Chopra, S. R. K.: The cranial suture closure in monkeys. *Proceedings of the Zoological Society of London* 128:67, 1957.

Chung, S. M. K., and Nissenbaum, M. M.: Congenital and developmental defects of the shoulder. *Orthopedic Clinics of North America* 6:381, 1975.

Claffey, T. J.: Avascular necrosis of the femoral head: an anatomical study. *Journal of Bone and Joint Surgery,* 42B:802, 1960.

Clanton, T. O., and DeLee, J. C.: Osteochondritis dissecans: history, pathophysiology and current treatment concepts. *Clinical Orthopaedics and Related Research* 167:50, 1982.

Cockburn, T. A.: The origin of the treponematoses. *Bulletin World Health Organization* 24:221, 1961.

Cockburn, A., and Cockburn, E.: *Mummies, Disease, and Ancient Cultures* (abridged edition). Cambridge, Cambridge University Press, 1980.

Cockshott, P. W.: Anatomical anomalies observed in radiographs of Nigerians—(I) Thoracic. *West African Medical Journal* 7(4):179, 1958.

Cohen, M. M., Jr.: History, terminology, and classification of craniosynostosis. In Cohen, M. M., Jr. (Ed): *Craniosynostosis: Diagnosis, Evaluation, and Management.* New York, Raven, 1986.

Coley, B.: *Neoplasms of Bone and Related Conditions.* New York, Paul B. Hoeber, 1949.

Congdon, R. T.: Spondylolisthesis and vertebral anomalies in skeletons of American aborigines. *Journal of Bone and Joint Surgery 14*B:511, 1931.

Cook, D. C.: Subsistence base and health in prehistoric Illinois Valley: evidence from the human skeleton. *Medical Anthropology 4:*109, 1979.

Cook, D. C., and Buikstra, J. E.: Health and differential survival in prehistoric populations: prenatal dental defects. *American Journal of Physical Anthropology 51:*649, 1979.

Correll, R. W., Jensen, J. L., and Rhyne, R. R.: Lingual cortical mandibular defects: a radiographic incidence study. *Oral Surgery Oral Medicine Oral Pathology 50*(3):287, 1980.

Cowell, M. J., and Cowell, H. R.: The incidence of spina bifida occulta in idiopathic scoliosis. *Clinical Orthopaedics and Related Research 118:*16, 1976.

Crelin, E. S.: Development of the musculoskeletal system. *Clinical Symposia 33*(1), 1981.

Cummings, S. R., Kelsey, J. L., Nevitt, M. C., and O'Dowd, K. J.: Epidemiology of osteoporosis and osteoporotic fractures. *Epidemiologic Review 7:*178, 1985.

Dahlin, D.: *Bone Tumors,* 2nd ed. Springfield, Thomas, 1967.

Dalinka, M. K., Dinnenberg, S., Greendyke, W. H., and Hopkins, R.: Roentgenographic features of osseous coccidioidomycosis and differential diagnosis. *Journal of Bone and Joint Surgery 53*A:1157, 1971.

Dallman, P., Siimes, R., and Stekel, A.: Iron deficiency in infancy and childhood. *American Journal of Clinical Nutrition 33:*86, 1980.

Dannels, E. G., and Nashel, D. J.: Periostitis: a manifestation of venous disease and skeletal hyperostosis. *Journal American Podiatric Association 73*(9):461, 1983.

Danzig, L. A., Greenway, G. and Resnick, D.: The Hill-Sachs lesion. *American Journal of Sports Medicine 8:*328, 1980.

Daveny, J. K., and Ross, M.D.: Cryptococcosis of bone. *African Journal of Medicine 15:*78, 1969.

David, D. J., Poswillo, D., and Simpson, D.: *The Craniosynostoses: Causes, Natural History and Management.* Berlin, Springer-Verlag, 1982.

David, R., Oria, R. A., Kumar, R., Singleton, E. B., Lindell, M. M., Shirkhoda, A., and Madewell, J. E.: Radiologic features of eosinophilic granuloma of bone. *American Journal of Roentgenology 153:*1021, 1989.

Davis, P. R.: The Thoraco-lumbar mortice joint. *Journal of Anatomy 89:*370, 1955.

Day, S. B.: Ossified subperiosteal hematoma. *Journal of American Medical Association 173:*84, 1960.

Derry, D. E.: Note on the accessory articular facets between the sacrum and ilium, and their significance. *Journal of Anatomy and Physiology 45:*202, 1911.

Dhooria, H. S., and Mody, R. N.: X-ray oddities. Unusually large ivory osteoma. *Journal Indian Dental Association 58*(3):89, 1986.

Dickel, D. N., and Doran, G. H.: Severe neural tube defect syndrome from the Early Archaic of Florida. *American Journal of Physical Anthropology 80:*325, 1989.

Dickson, R. A.: Conservative treatment for idiopathic scoliosis. *Journal of Bone and Joint Surgery 67*B:176, 1985.

Dieppe, P. A., Bacon, P. A., Bamji, A. N., and Watt, I.: *Atlas of Clinical Rheumatology.* Philadelphia, Lea & Febiger, 1986.

Dugdale, A. E., Lewis, A. N., and Canty, A. A.: The natural history of chronic otitis media. *New England Journal of Medicine 307:*1459, 1982.

Dutour, O.: Enthesopathies (lesions of muscular insertions) as indicators of the activities of neolithic Saharan populations. *American Journal of Physical Anthropology 71:*221, 1986.

Echols, R. M., Selinger, D. D., Hallowell, C., Goodwin, J. S., Duncan, M. H., and Cushing, A. H.: Rhizopus osteomyelitis: a case report and review. *American Journal of Medicine 66:*141, 1979.

Edelson, J. G., Nathan, H., and Arensburg, B.: Diastematomyelia—the "double-barrelled" spine. *Journal of Bone and Joint Surgery 69*B(2):188, 1987.

Edwards, W. G.: Complications of suppurative otitis media. In Ludman, H. (Ed.): *Mawson's Diseases of the Ear,* 5th ed. Chicago, Year Book Medical Publishers, Inc., 1988.

Eichenholtz, S. N.: *Charcot Joints.* Springfield, Thomas, 1966.

Eisenstein, S.: The morphometry and pathological anatomy of the lumbar spine in South African Negroes and Caucasoids with specific reference to spinal stenosis. *Journal of Bone and Joint Surgery 59*B:173, 1977.

——: Spondylolysis: a skeletal investigation of two populations. *Journal of Bone and Joint Surgery 60*B:488, 1978.

Ell, S. R.: Reconstructing the epidemiology of medieval leprosy: preliminary efforts with regard to Scandinavia. *Perspectives in Biology and Medicine 31:*496, 1988.

El-Najjar, M. Y., Ryan, D. J., Turner, C. G., and Lozoff, B.: The etiology of porotic hyperostosis among the prehistoric and historic Anasazi Indians of southwestern United States. *American Journal of Physical Anthropology 44:*477–488, 1976.

Epstein, B. S.: The concurrence of parietal thinness with postmenopausal, senile or idiopathic osteoporosis. *Radiology 60:*29, 1953.

Epstein, H. C.: Traumatic dislocations of the hip. *Clinical Orthopaedics and Related Research 92:*116, 1973.

——: *The Spine: A Radiological Text and Atlas,* 4th Ed. Philadelphia, Lea and Febiger, 1976.

Farkas, A.: Physiological scoliosis. *Journal of Bone and Joint Surgery 23*(3):607, 1941.

Ferembach, D.: Frequency of spina bifida occulta in prehistoric human skeletons. *Nature 199:*100, 1963.

Fiese, M. J.: *Coccidioidomycosis.* Springfield, Thomas, 1958.

Fink, I. J., Pastakia, B., and Barranger, J. A.: Enlarged phalangeal nutrient foramina in Gaucher disease and B-thalassemia major. *American Journal of Roentgenology 143:*647, 1984.

Finnegan, M., and Faust, M. A.: *Bibliography of Human and Nonhuman Non-metric*

Variation. Research Reports No. 14, Department of Anthropology, University of Massachusetts. Amherst, University of Massachusetts, 1974.

Fiumara, N. J., and Lessell, S.: Manifestations of late congenital syphilis: an analysis of 271 patients. *Archives of Dermatology 102:*78, 1970.

——: The stigmata of late congenital syphilis: an analysis of 100 patients. *Sexually Transmitted Diseases 10*(3):126, 1983.

Floman, Y., Bloom, R. A., and Robin, G. C.: Spondylolysis in the upper lumbar spine: a study of 32 patients. *Journal of Bone and Joint Surgery 69*B(4):582, 1987.

Francois, R. J.: Microradiographic study of the intervertebral bridges in ankylosing spondylitis and in the normal sacrum. *Annals of Rheumatic Diseases 24:*481, 1965.

Fredrickson, B. E., Baker, D., McHolick, W. J., Yuan, H. A., and Lubicky, J. P.: The natural history of spondylolysis and spondylolisthesis. *Journal of Bone and Joint Surgery 66*A:699, 1984.

Gardner, D. L.: *Pathology of the Connective Tissue Diseases.* London, Edward Arnold, 1965.

Garth, W. P., and Van Patten, P. K.: Fractures of the lumbar lamina with epidural hematoma simulating herniation of a disc: a case report. *Journal of Bone and Joint Surgery 71*A:771, 1989.

Genant, H. K.: Osteoporosis and bone mineral assessment. In McCarty, D. J. (Ed.): *Arthritis and Allied Conditions: A Textbook of Rheumatology,* 11th ed. Philadelphia, Lea and Febiger, 1989.

Genner, Byron A.: Fracture of the supracondyloid process. *Journal of Bone and Joint Surgery 41*A:1333, 1959.

Gensburg, R. S., Kawashima, A., and Sandler, C. M.: Scintigraphic demonstration of lower extremity periostitis secondary to venous insufficiency. *Journal of Nuclear Medicine 29*(7):1279, 1988.

George, G. R.: Bilateral bipartite patellae. *British Journal of Surgery 22:*555, 1935.

Gibson, M. J., Szypryt, E. P., Buckley, J. H., Worthington, B. S., and Mulholland, R. C.: Magnetic resonance imaging of adolescent disc herniation. *Journal of Bone and Joint Surgery 69*B(5):703, 1987.

Giroux, J. C., and Leclerq, T. A.: Lumbar disc excision in the second decade. *Spine 7:*168, 1982.

Goldsand, G.: Actinomycosis. In Hoeprich, P. D. and Jordan, M. C. (Eds.): *Infectious Disease: A Modern Treatise of Infectious Processes,* 4th Ed. Philadelphia, Lippincott, 1989, pp. 666–684.

Goldsmith, W. M.: The Catlin Mark: the inheritance of an unusual opening in the parietal bones. *Journal of Heredity 13:*69, 1922.

Goodman, A. H., and Armelagos, G. J.: Childhood stress and decreased longevity in a prehistoric population. *American Anthropologist 90*(4):936, 1988.

Gorab, G. N., Brahney, C., and Aria, A. A.: Unusual presentation of a Stafne bone cyst. *Oral Surgery Oral Medicine Oral Pathology 61*(3):213, 1986.

Grant, J. C. B.: *A Method of Anatomy: Descriptive and Deductive,* 4th ed. Baltimore, Williams and Wilkins, 1948.

——: *Grant's Atlas of Anatomy,* 6th ed. Baltimore, Williams and Wilkins, 1972.

Graves, W. W.: Observations on age changes in the scapula. *American Journal of Physical Anthropology 5:*21, 1922.

Gray, H.: *Anatomy of the Human Body,* 25th ed. Philadelphia, Lea and Febiger, 1948.

Green, W. T.: Painful bipartite patellae: a report of three cases. *Clinical Orthopaedics and Related Research 110:*197, 1975.

Greer, A. E.: *Disseminating Fungal Diseases of the Lung.* Springfield, Thomas, 1962.

Gregg, J. B., and Gregg, P. S.: *Dry Bones: Dakota Territory Reflected.* Sioux Falls, Sioux Falls Printing, 1987.

Griffiths, H. J.: *Basic Bone Pathology.* Norwalk, Appleton-Century-Crofts, 1981.

Gruber, W.: Ueber einen neuen sekundarrn Tarsalknochen -Calcaneus secundarius — mit Bemerkungen uber den Tarsus uberhaupt. *Memoirs de l'Academie des Science de St. Petersbourg, T. 17,* No. 6, 1871.

Grupe, G.: Metastasizing carcinoma in a medieval skeleton: differential diagnosis and etiology. *American Journal of Physical Anthropology 75:*369, 1988.

Hackett, C. J.: The human treponematoses. In Brothwell, D. R. and Sandison, A. T. (Eds.): *Diseases in Antiquity: A Survey of the Diseases Injuries and Surgery of Early Populations.* Springfield, Thomas, 1967, pp. 152–169.

———: Diagnostic criteria of syphilis, yaws and treponarid (treponematoses) and some other diseases in dry bone. *Sitzunosbericht der heidelberger Akademie der Wissenschaft 4,* 1976.

———: Development of caries sicca in a dry calvaria. *Virchows Archives (Pathol Anat) 391:*53, 1981.

Haibach, H., Farrell, C., and Gaines, R. W.: Osteoid osteoma of the spine: surgically correctable cause of painful scoliosis. *Canadian Medical Association Journal 135*(8):895, 1986.

Haidar, Z., and Kalamchi, S.: Painful dysphagia due to fracture of the styloid process. *Oral Surgery Oral Medicine Oral Pathology 49*(1):5, 1980.

Halpern, A. A., and Hewitt, O.: Painful medial bipartite patellae: a case report. *Clinical Orthopaedics and Related Research 134:*180, 1978.

Handler, J. S., Corruccini, R. S., and Mutaw, R. J.: Tooth mutilation in the Caribbean: evidence from a slave burial population in Barbados. *Journal of Human Evolution 11:*297, 1982.

Hansen, G. L.: Pers. Comm., Smithsonian Institution, Washington.

Hansson, L. G.: Development of a lingual mandibular bone cavity in an 11-year-old boy. *Oral Surgery Oral Medicine Oral Pathology 49*(4):376, 1980.

Harris, R. I., and Wiley, J. J.: Acquired spondylolysis as a sequel to spine fusion. *Journal of Bone and Joint Surgery 45*A:1159, 1963.

Harvey, W., and Noble, H. W.: Defects on the lingual surface of the mandible near the angle. *British Journal Oral Surgery 6:*75, 1968–69.

Hellman, M. H.: Charcot joint disease (Charcot joints). In McCarty, D. J. (Ed.): *Arthritis and Allied Conditions: A Textbook of Rheumatology,* 11th ed. Philadelphia, Lea and Febiger, 1989.

Hengen, O. P.: Cribra orbitalia: pathogenesis and probable etiology. *HOMO 22:*57, 1971.

Henrard, J. C., and Bennett, P.H.: Etude epidemiologique de l'hyperostose vertebrale.

Enquete dans une population adulte d'Indiens d'Amerique. *Revue du Thumatisme et des Maladies Osteoarticulaires 40:*581, 1973.

Hertzler, A. E.: Surgical pathology of the diseases of bone. *Hertzler's Monographs on Surgical Pathology,* Chicago, Lakeside, 1931.

Hilel, N.: The para-articular processes of the thoracic vertebrae. *Journal of Anatomy 133:*605, 1959.

——: Osteophytes of the vertebral column: an anatomical study of their development according to age, race, and sex with considerations as to their etiology and significance. *Journal of Bone and Joint Surgery 44*A:243, 1962.

Hill, M. C.: Porotic hyperostosis: a question of correlations verses causality. *American Journal of Physical Anthropology 66:*182, 1985.

Hirata, K.: A contribution to the paleopathology of cribra orbitalia in Japanese. 1. Cribra orbitalia in Edo Japanese. *St. Marianna Medical Journal 16:*6, 1988.

Hitchcock, H. H.: Spondylolisthesis: observations on its development, progression, and genesis. *Journal of Bone and Joint Surgery 22*B:1, 1940.

Hoeprich, P. D.: Coccidioidomycosis. In Hoeprich, P. D. and M. C. Jordan (Eds.): *Infectious Diseases: A Modern Treatise of Infectious Processes,* 4th Ed. Philadelphia, Lippincott, 1989, pp. 489–502.

——: Nonspecific treponematoses. In Hoeprich, P. D. and M. C. Jordan (Eds.): *Infectious Diseases: A Modern Treatise of Infectious Processes,* 4th Ed. Philadelphia, Lippincott, 1989, pp. 1021–1034.

Hoffman, J. M.: Enlarged parietal foramina—their morphological variation and use in assessing prehistoric biological relationships. In Hoffman, J. M. and Brunker, L. (Eds.): *Studies in California Paleopathology,* Contributions of the University of California Archaeological Research Facility, No. 30. Berkeley, University of California, 1976.

Holcomb, R. C.: *Who Gave the World Syphilis? The Haitian Myth.* New York, Froben, 1930.

——: The antiquity of congenital syphilis. *Medical Life 42:*275, 1935.

Hooton, E. A.: *The Indians of Pecos Pueblo, A study of their skeletal remains.* Papers of the Southwestern Expedition No. 4. New Haven, Yale University Press, 1930.

Hoppenfeld, S.: Back pain. *Pediatric Clinics of North America 24*(4):881, 1989.

Horal, J., Nachemson, A., and Scheller, S.: Clinical and radiological long term follow-up of vertebral fractures in children. *Acta Orthopedia Scandinavia 43:*491, 1972.

Hough, A. J., and Sokoloff, L.: Pathology of osteoarthritis. In McCarty, D. J. (Ed.): *Arthritis and Allied Conditions: A Textbook of Rheumatology,* 11th ed. Philadelphia, Lea and Febiger, 1989.

Houghton, P.: The relationship of the pre-auricular groove of the ilium to pregnancy. *American Journal of Physical Anthropology 41:*381, 1974.

——: The bony imprint of pregnancy. *Bulletin of the New York Academy of Medicine 51:*655, 1975.

Hrdlicka, A.: *Anthropological Work in Peru in 1913 With Notes on the Pathology of the Ancient Peruvians.* Washington, Smithsonian Institution, 1914.

Hudson, E. H.: Yaws and syphilis; same or different? *Navy Medical Bulletin 38:*172, 1958a.

——: *Non-Venereal Syphilis: A sociological and medical study of bejel.* Edinburgh, E. & S. Livingstone, 1958b.

Hughston, J. C., Hergenroeder, P. J., and Courtenay, B. G.: Osteochondritis dissecans of the femoral condyles. *Journal of Bone and Joint Surgery 66*A(9):1340, 1984.

Hutchinson, D. L., and Larsen, C. S.: Determination of stress episode duration from linear enamel hypoplasias: a case study from St. Catherine's Island, Georgia. *Human Biology 60:*93, 1988.

Hutchinson, J.: *A Clinical Memoir of Certain Diseases of the Eye and Ear Consequent on Inherited Syphilis.* Journal of Churchill, London, 1863.

Hutter, R., Worcester, J., Francis, K., Foote, F., and Stewart, F.: Benign and malignant giant cell tumors of bone. *Cancer 15:*653, 1962.

Ihle, C. L., and Cochran, R. M.: Fracture of the fused os trigonum. *American Journal Sports Medicine 10*(1):47, 1982.

Jaffe, H. L.: *Tumors and Tumorous Conditions of Bones and Joints.* Philadelphia Lea and Febiger, 1958.

——: Ischemic necrosis of bone. *Medical Radiography and Photography 45:*58, 1969.

——: *Metabolic, Degenerative and Inflammatory Diseases of Bone and Joints.* Philadelphia, Lea and Febiger, 1975.

Jit, I., and Kaur, H.: Rhomboid fossa in the clavicles of North Indians. *American Journal of Physical Anthropology 70:*97, 1986.

Johansson, L. U.: Bone and related materials. In Hodge, W. M. (Ed.): *In situ Archaeological Conservation.* Mexico, Instituto Nacional de Antropologia e Historia, 1987.

Jones, R. R., and Martin, D. D.: Blastomycosis of bone. *Surgery 10:*931, 1941.

Julkunen, H., Heinonen, O. P., Knekt, P. and Maatela, J.: The epidemiology of hyperostosis of the spine together with its symptoms and related mortality in a general population. *Scandinavian Journal of Rheumatology 40:*581, 1973.

Karpini, M. R. K., Newton, G., and Henry, A. P. J.: The results and morbidity of varus osteotomy for Perthes disease. In Burwell, R. G. and Harrison, M. H. M. (Eds.): *Clinical Orthopaedics and Related Research 209:30, 1986.*

Kate, B. R.: The incidence and cause of cervical fossa in Indian femora. *Journal of the Anatomical Society of India 12*(2):69, 1963.

Keats, T. E.: *An Atlas of Normal Roentgen Variants.* Chicago, Year Book, 1973.

Keim, Hugo A.: Scoliosis. *Clinical Symposia 24,* No. 1, 1972.

——: Low back pain. *Clinical Symposia 25,* No. 3, 1973.

Kelley, M. A.: Pers. Comm., University of Rhode Island.

Kelley, M. A., and Eisenberg, L. E.: Blastomycosis and tuberculosis in early American Indians: a biocultural view. *Midcontinental Journal of Archaeology 12:*89, 1987.

Kelley, M. A., and El-Najjar, M. V.: Natural variation and differential diagnosis of skeletal changes in tuberculosis. *American Journal of Physical Anthropology 52:*153, 1980.

Kelley, M. A., and Micozzi, M. S.: Rib lesions in chronic pulmonary tuberculosis. *American Journal of Physical Anthropology 65:*381, 1984.

Kellgren, J. J., and Lawrence, J. S.: Osteoarthritis and disk degeneration in an urban population. *Annals Rheumatic Diseases 17:*388, 1958.

Kennedy, G. E.: The relationship between auditory exostosis and cold water: a latitudinal analysis. *American Journal of Physical Anthropology 71:*401, 1986.

Keur, J. J., Campbell, J. P. S., McCarthy, J. F., and Ralph, W. J.: The clinical significance of the elongated styloid process. *Oral Surgery Oral Medicine Oral Pathology 61:*399, 1986.

Khazhinskaia, V. A. and Ginzburg, M. A.: X-ray anatomic variants of the rhomboid fossa of the clavicle. *Vestnik Rentgenolog Radiologia 3:*32, 1975.

Klippel, M., and Feil, A.: Un cas d'absence des vertebres cervicales. *Nouv Iconong Salpetriere 25:*223, 1912.

Knowles, A. K.: Acute traumatic lesions. In Hart, G. D. (Ed.): *Disease in Ancient Man: An International Symposium.* Ontario, Irwin, 1983.

Kostick, E. L.: Facets and imprints on the upper and lower extremities of femora from a Western Nigerian population. *Journal of Anatomy 97*(3):393, 1963.

Krogman, W. M.: The pathologies of pre- and protohistoric man. *Ciba Symposia 2:*432, 1940.

Kromberg, J. G., and Jenkins, T.: Common birth defects in South African blacks. *South African Medical Journal 62:*599, 1982.

Krugman, S., Katz, S. L., Gershon, A. A., and Wilfert, C.: *Infectious Diseases of Children,* 8th ed. St. Louis, Mosby, 1985.

Kulkarni, V. N., and Mehta, J. M.: Tarsal disintegration (TD) in leprosy. *Leprosy in India 55:*338, 1983.

Kullmann, L., and Wouters, H. W.: Neurofibromatosis, gigantism and subperiosteal haematoma: report of two children with extensive subperiosteal bone formation. *Journal of Bone and Joint Surgery 54*A:130, 1972.

Ladisch, S., and Jaffe, E. S.: The histocytoses. In Pizzo, P. A. and Poplak, D. G. (Eds.): *Principles and Practice of Pediatric Oncology.* Philadelphia, Lippincott, 1989.

Lallo, J. W., Armelagos, G. J., and Mensforth, R. P.: The role of diet, disease and physiology in the origin of porotic hyperostosis. *Human Biology 48*(3):47I, 1977.

Lamy, C., Bazergui, A., Kraus, H., and Farfan, H. F.: The strength of the neural arch and the etiology of spondylolysis. *Orthopedic Clinics of North America 6:*215, 1975.

Lane, N. E., Bloch, D. A., Jones, H. H., Simpson, U., and Fries, J. F.: Osteoarthritis in the hand: a comparison of handedness and hand use. *Journal of Rheumatology 16*(5):637, 1989.

Lane, N. E., Bloch, D. A., Jones, H. H., Marshall, W. H., Wood, P. D., and Fries, J. F.: Long-distance running, bone density and osteoarthritis. *Journal of American Medical Association 255:*1147, 1986.

Lange, R. H., Lange, T. A., and Rao, B. K.: Correlative radiographic, scintigraphic, and histological evaluation of exostoses. *Journal of Bone and Joint Surgery 66*A(9):1454, 1984.

Langlais, R. P., Miles, D. A., and Van Dis, M. L.: Elongated and mineralized stylohyoid ligament complex: a proposed classification and report of a case of Eagle's syndrome. *Oral Surgery Oral Medicine Oral Pathology 61:*527, 1986.

Lanzkowsky, P.: Radiological features of iron deficiency anemia. *American Journal of Diseases of Children 116:*16, 1968.

Latham, R. A., and Burston, W. R.: The postnatal pattern of growth at the sutures of the human skull. *Dental Practitioner 17:*61, 1966.

Laurent, L. E., and Einola, S.: Spondylolisthesis in children and adolescents. *Acta Orthopedica Scandinavica 31:*45–64, 1961.

Lawrence, J. S.: Rheumatism in cotton cooperative. *British Journal Indian Medicine 18:*270, 1961.

———: Generalized osteoarthritis in a population sample. *American Journal Epidemiology 90:*381, 1969.

Lester, C. W., and Shapiro, H. L.: Lumbar vertebrae of pre-historic American Eskimos: a study of skeletons in the American Museum of Natural History, chiefly from Point Hope, Alaska. *American Journal of Physical Anthropology 28:*43, 1968.

Lestini, W. F., and Wiesel, S. W.: The pathogenesis of cervical spondylosis. *Clinical Orthopaedics and Related Research 239:*69, 1989.

Letts, M., Smallman, T., Afanasiev, R., and Gouw, G.: Fracture of the pars interarticularis in adolescent athletes: a clinical-biomechanical analysis. *Journal of Pediatric Orthopedics 6:*40, 1986.

Liberson, F.: Os acromiale—a contested anomaly. *Journal of Bone and Joint Surgery 19:*683, 1937.

Libson, E., Bloom, R. A., and Dinari, G.: Symptomatic and asymptomatic spondylolysis and spondylolisthesis in young adults. *International Orthopedics 6:*259, 1982.

Lichenstein, L.: Histiocytosis X. Integration of eosinophilic granuloma of bone, Letterer-Siwe disease, and Schuller-Christian disease as related manifestations of a single nosologic entity. *Archives of Pathology 56:*84, 1953.

———: *Diseases of Bone and Joints.* Saint Louis, Mosby, 1970.

Lima, Mauricio D. L. P.: *Contribucicao ao Estudo do Os Trigonum Tarsi.* San Paulo, These Inaugural, 1928.

Limson, M.: Metopism as found in Filipino skulls. *American Journal of Physical Anthropology 7(3):*317, 1924.

Lindblom, K.: Bachache and its relation to ruptures of the intravertebral disks. *Radiology 57:*710, 1951.

Lisowski, F. P.: Prehistoric and early historic trepanation. In Brothwell, D. R. and Sandison, A. T. (Eds.): *Diseases in Antiquity: A Survey of Diseases Injuries and Surgery of Early Populations.* Springfield, Thomas, 1967.

Lodge, T.: Thinning of the parietal bones in early Egyptian populations and its aetiology in the light of modern observations. In Brothwell, D. R. and Sandison, A. T. (Eds.): *Diseases in Antiquity: A Survey of Diseases Injuries and Surgery of Early Populations.* Springfield, Thomas, 1967.

Lonstein, M., and Hochschuler, H.: Differential diagnosis of lumbar disc disease. *Spine 3(1):*57, 1989.

Lundy, J. K.: A report on the use of Fully's anatomical method to estimate stature in military skeletal remains. *Journal of Forensic Sciences 33(2):*534, 1988.

MacAusland, W. R., and Mayo, R. A.: *Orthopedics: A Concise Guide to Clinical Practices.* Boston, Little, Brown, 1965.

Madewell, J. E., Ragsdale, B. D., and Sweet, D. E.: Radiologic and pathologic analysis of solitary bone lesions. Part I: internal margins. *Radiologic Clinics of North America 19*(4):715, 1981.

Mahoubi, S.: CT appearance of nidus in osteoid osteoma versus sequestration in osteomyelitis. *Journal Computer Assisted Tomography 10*(3):457, 1986.

Manchester, K.: *The Archaeology of Disease.* Bradford, University of Bradford, 1983.

——: Bone changes in leprosy: pathogenesis and paleopathology. Paper presented at the Annual Meeting of the Association of Physical Anthropologists, San Diego, 1989.

Mann, R. W.: Calcaneus secundarius: description of a common accessory ossicle. *Journal of American Podiatric Medical Association 79*(8):363, 1989.

Mann, R. W., Meadows, L., Bass, W. M., and Watters, D. R.: Description of skeletal remains from a black slave cemetery from Montserrat, West Indies. *Annals of Carnegie Museum 56:*319, 1987.

Mann, R. W., and Murphy, S. P.: Skeletal indicators of physical stress in soldiers from the Battle of Fort Erie (War of 1812), Canada. Paper presented at the Annual Meeting of the Northeastern Anthropological Association, Montreal, Canada, 1989.

Mankin, H. J.: Rickets, osteomalacia, and renal osteodystrophy, part I. *Journal Bone and Joint Surgery 56*A:101, 1974.

Manzanares, M. C., Goret-Nicaise, M., and Dhem, A.: Metopic sutural closure in the human skull. *Journal of Anatomy 161:*203, 1988.

Markell, E. K., and Voge, M.: *Medical Parasitology* 5th ed. Philadelphia, Saunders, 1981.

Markowitz, H. A., and Gerry, R. G.: Temporomandibular joint disease. *Oral Surgery, Oral Medicine, Oral Pathology 3:*75, 1950.

Masi, A. T., and Medsger, T. A.: Epidemiology of the rheumatic diseases. In McCarty, D.J. (Ed.): *Arthritis and Allied Conditions: A Textbook of Rheumatology.* 11th ed. Philadelphia, Lea and Febiger, 1989.

Matossian, R. M., Richard, M. D., and Smyth, J. D.: Hydatidosis: a global problem of increasing importance. *Bulletin World Health Organization 55:*499, 1977.

McAuliffe, T. B., Hilliar, K. M., Coates, C. J., and Grange, W. J.: Early immobilization of Colles' fractures: a retrospective trial. *Journal of Bone and Joint Surgery 69*B(5):727, 1987.

McKee, B. W., Alexander, W. J., and Dinbar, J. S.: Spondylolysis and spondylolisthesis in children. *Journal Canadian Association Radiology 22:*100, 1971.

McKern, T. W., and Stewart, T.D.: *Skeletal Age Changes in Young American Males: Analyzed from the standpoint of age identification.* Technical Report EP-45, Quartermaster Research and Development Center, Massachusetts, 1957.

Meadows, L.: Pers. Comm., University of Tennessee, Knoxville.

Meisel, A. D., and Bullough, P. G.: *Atlas of Osteoarthritis.* New York, Lea and Febiger, 1984.

Mensforth, R. P., Lovejoy, C. O., Lallo, J. W., and Armelagos, G. J.: The role of constitutional factors, diet, and infectious disease in the etiology of porotic hyperostosis and periosteal reactions in prehistoric infants and children. *Medical Anthropology 2* (Winter), part 2, 1978.

Merbs, C. F.: *Patterns of Activity-Induced Pathology in A Canadian Inuit Population.* Archaeological Survey of Canada Paper, Mercury Series 119. Ottawa, National Museum of Man, 1983.

Merbs, C. F., and Euler, R. C.: Atlanto-occipital fusion and spondylolisthesis in an Anasazi skeleton from Bright Angel Ruin, Grand Canyon National Park, Arizona. *American Journal of Physical Anthropology 67:*381, 1985.

Micozzi, M. S., and Kelley, M. A.: Evidence for precolumbian tuberculosis at the Point of Pines Site, Arizona: skeletal pathology in the sacro-iliac region. In Merbs, C.F. and Miller, R.J. (Eds.): *Health and Disease in the Prehistoric Southwest.* Anthropological Research Papers No. 34. Tempe, Arizona State University Press, 1985.

Miles, J. S.: *Orthopedic Problems of the Wetherill Mesa Populations.* Washington, D.C., Publications in Archaeology 7G, National Park Service, 1975.

Miller, M. W., and Keagy, R. M.: Enlarged parietal foramina. Eight examples in four generations. *Medical Radiography and Photography 32:*74, 1956.

Mintz, G., and Fraga, A.: Severe osteoarthritis of the elbow in foundry workers. *Archives Environment Health 27:*78, 1973.

Mirra, J. M.: Pathogenesis of Paget's disease based on viral etiology. *Clinical Orthopaedics and Related Research 217:*162, 1987.

Moe, J. H., Winter, R. B., Bradford, D. S., and Lonstein, J. E.: *Scoliosis and Other Spinal Deformities.* Philadelphia, Saunders, 1978.

Moller-Christensen, V.: *Ten lepers from Naestved in Denmark: a study of skeletons from a medieval Danish leper hospital.* Copenhagen, Danish Science Press, 1953.

———: *Bone Changes in Leprosy.* Copenhagen, Munksgaard, 1961.

———: Evidence of Leprosy in Earlier Peoples. In Brothwell, D. R. and Sandison, A. T. (Eds.): *Diseases in Antiquity: A Survey of the Diseases Injuries and Surgery of Early Populations.* Springfield, Thomas, 1967.

———: Changes in the anterior nasal spine and alveolar process of maxilla in leprosy: a clinical examination. *International Journal of Leprosy 42:*431, 1974.

———: *Leprosy changes in the skull.* Odense, Odense University Press, 1978.

———: Leprosy and tuberculosis. In Hart, G.D. (Ed.): *Disease in Ancient Man: An International Symposium.* Ontario, Irwin, 1983.

Moller-Christensen, V., and Sandison, A.T.: Ursa orbitae (cribra orbitalia) in the collection of crania in the anatomy department of the University of Glascow. *Pathologia et Microbiologia 26:*175, 1963.

Monsour, P. A., and Young, W. G.: Variability of the styloid process and stylohyoid ligament in panoramic radiographs. *Oral Surgery Oral Medicine Oral Pathology 61:*522, 1986.

Moore, K. L.: *The Developing Human: Clinically Oriented Embryology.* London, Saunders, 1982.

Moore, R. M., and Green, N. E.: Blastomycosis of bone: a report of six cases. *Journal of Bone and Joint Surgery 64*A(7):1097, 1982.

Moore, S.: *Hyperostosis Cranii.* Springfield, Thomas, 1955.

Morse, D.: Prehistoric tuberculosis in America. *American Review of Respiratory Disease 83:*489, 1961.

——: *Ancient Disease in the Midwest.* Illinois State Museum Reports of Investigations, No. 15, 1969.

Morton, N. E.: Birth defects in racial crosses. In *Congenital Malformation: Proceedings of the Third International Conference.* New York, Excerpta Medica, 1970.

Moseley, J. E.: *Bone Changes in Hematologic Disorders.* New York, Grune and Stratton, 1963.

Moskowitz, R. W., Howell, D. S., Goldberg, V. M., and Mankin, H. J.: *Osteoarthritis: Diagnosis and Management.* Philadelphia, Saunders, 1984.

Mudge, K., Wood, V. E., and Frykman, G. K.: Rotator cuff tears associated with os acromiale. *Journal of Bone and Joint Surgery 66*A(3):427, 1984.

Muller, F., O'Rahilly, and Benson, D. R.: The early origin of vertebral anomalies, as illustrated by a "butterfly vertebra". *Journal of Anatomy 149:*157, 1986.

Murray, J. F., Merriweather, A. M., and Freedman, M. L.: Endemic syphilis in the Bakwena Reserve of the Bechuanaland protectorate. *Bulletin of World Health Organization 15:*975, 1956.

Murray, K.: Pers. Comm., Smithsonian Institution, Washington.

Naffsiger, H. C., Inman, V., and Saunders, J.: Lesions of intervertebral disc and ligamenta flava: clinical and anatomical studies. *Surgery, Gynecology and Obstetrics 66:*288, 1938.

Nathan, H.: Spondylolysis: its anatomy and mechanism of development. *Journal of Bone and Joint Surgery 41*A:303, 1959.

Newman, P. H.: The etiology of spondylolysis. *Journal of Bone and Joint Surgery 45*B(1):39–59, 1963.

Niswander, J. D., Barrow, M. V., and Bingle, G. J.: Congenital malformations in the American Indian. *Social Biology 22*(3):203, 1975.

Noble, E. R., and Noble, G. A.: *Parasitology: The Biology of Animal Parasites* 5th ed. Philadelphia, Lea and Febiger, 1982.

Norman, A.: Roentgenologic diagnosis. In Moskowitz, R. W., Howell, D. S., Goldberg, V. M., and Mankin, H. J. (Eds.): *Osteoarthritis: Diagnosis and Management.* Philadelphia, Saunders, 1984.

O'Duffy, J. D.: Spinal Stenosis. In McCarty, D.J. (Ed.): *Arthritis and Allied Conditions: A Textbook of Rheumatology,* 11th ed. Philadelphia, Lea and Febiger, 1989.

Ogden, J. A., McCarthy, S. M., and Jokl, P.: The painful bipartite patella. *Journal of Pediatric Orthopedics 2*(3):263, 1982.

Olmsted, W. M.: Some skeletogenic lesions with common calvarial manifestations. *Radiologic Clinics of North America 19:*703, 1981.

Ortner, D. J.: Descriptions and classifications of degenerative bone changes in the distal joint surfaces of the humerus. *American Journal of Physical Anthropology 28:*139, 1968.

Ortner, D. J., and Putschar, W. G. J.: *Identification of Pathological Conditions in Human Skeletal Remains.* Washington, Smithsonian Institution Press, 1985.

Owsley, D. W., Orser, C. E., Mann, R. W., Moore-Jansen, P. H., and Montgomery,

R. L.: Demography and pathology of an urban slave population from New Orleans. *American Journal of Physical Anthropology 74:*185, 1987.

Paget, J.: On the production of some of the loose bodies in the joints. *St. Bartholomew's Hospital Report 6:*1, 1870.

———: On a form of chronic inflammation of bones (osteitis deformans). *Transactions of the Medical Chirurgical Society 60:*37, 1877.

Palkovich, A. M.: Endemic disease patterns in paleopathology: porotic hyperostosis. *American Journal of Physical Anthropology 74:*526, 1987.

Panush, R. S., Schmidt, C., Caldwell, J. R., et al.: Is running associated with degenerative joint disease? *Journal of American Medical Association 255:*1152, 1986.

Paparella, M. M., Goycoolea, M. V., and Meyerhoff, W. L.: Inner ear pathology and otitis media: A review. *Annals of Otology. Rhinology and Laryngology 89*(68):249, 1980.

Parsons, F. G.: On the proportions and characteristics of the modern English clavicle. *Journal of Anatomy 51:*71, 1916.

Paterson, D. E.: *Bones changes in leprosy.* Leprosy in India *28:*128, 1956.

Paterson, D. E., and Job, C. K.: Bone changes and absorption in leprosy. In Cochrane, R. G. and T. F. Davey (Eds.): *Leprosy in Theory and Practice.* Bristol, John Wright and Sons, 1964.

Pauwels, F.: *Biomechanics of the Normal and Diseased Hip.* Berlin, Springer-Verlag, 1976.

Pavithran, K.: Acquired syphilis in a patient with late congenital syphilis. *Sexually Transmitted Diseases 14*(2):119, 1987.

Pindborg, J. J.: *Pathology of the Dental Hard Tissue.* Philadelphia, Saunders, 1970.

Pitt, M. J.: Rachitic and osteomalacic syndromes. *Radiologic Clinics of North America 19*(4):581, 1981.

Pochaczevsky, R., Yen, Y. M., and Sherman, R. S.: The roentgen appearance of benign osteoblastoma. *Radiology 75:*429, 1960.

Ponseti, Ignacio V., El-Khoury, G. Y., Ippolito, E., and Weinstein, S. L.: A radiographic study of skeletal deformities in treated clubfeet. *Clinical Orthopaedics and Related Research 160:*30, 1981.

Poirier, P.: *Traité d'anatomie humaine I:*515. Poireir & Charpy, Paris, 1911.

Poswall, B. D.: Coccidioidomycosis and North American blastomycosis: differential diagnosis of bone lesions in pre-Columbian American Indians. *American Journal of Physical Anthropology 44:*199, 1976.

Procknow, J. J., and Loosli, C. G.: Treatment of the deep mycosis. *American Medical Association Archives of Internal Medicine 101:*765, 1958.

Prokopec, M., Simpson, D., Morris, L., and Pretty, G.: Craniosynostosis in a prehistoric aboriginal skull: a case report. *Ossa 9/11:*111, 1982–1984.

Prusick, V. R., Samberg, L. C., and Wesolowski, D. P.: Klippel-Feil syndrome associated with spinal stenosis: a case report. *Journal of Bone and Joint Surgery 67*A:161, 1985.

Puramen, J., Ala-Ketola, L., Pelto Kallio, P.: Running and primary osteoarthritis of the hip. *British Medical Journal 2:*424, 1975.

Pusey, W. A.: The beginning of syphilis. *Journal of American Medical Association* *44:*1961, 1915.

Raisz, L. G.: Osteoporosis. *Journal of the American Geriatrics Society 30*(2):127, 1982.

Ravichandran, G.: Upper lumbar spondylolysis. *International Orthopedics 5:*31, 1981.

Ray, R., Jr.: *Catalogue of the Pathological Cabinet of the New York Hospital.* New York, S & W Wood, 1860.

Reeves, R. J., and Pederson, R.: Fungus infection of bones. *Radiology 62:*55, 1954.

Reichart, P.: Facial and oral manifestations in leprosy. *Oral Surgery Oral Medicine Oral Pathology 41*(3):385, 1976.

Resnick, D., and Greenway, G.: Distal femoral cortical defects, irregularities, and excavations. *Radiology 143*(2):345, 1982.

Resnick, D., and Niwayama, G.: Radiographic and pathologic features of spinal involvement in diffuse idiopathic skeletal hyperostosis (DISH). *Radiology 119*(3): 559, 1976.

——: *Diagnosis of Bone and Joint Disorders.* Philadelphia, Saunders, 1981.

——: Anatomy of individual joints. In Resnick, D. and Niwayama, G. (Eds.): *Diagnosis of Bone and Joint Disorders.* Philadelphia, Saunders, 1981.

——: *Diagnosis of Bone and Joint Disorders,* 2nd ed. Philadelphia, Saunders, 1988.

Richards, L. C.: Temporomandibular joint morphology in two Australian aboriginal populations. *Journal Dental Research 66:*1602, 1987.

——: Degenerative changes in the temporomandibular joint in two Australian aboriginal populations. *Journal Dental Research 67:*1529, 1988.

Richards, L. C., and Brown, T.: Dental attrition and degenerative arthritis of the temporomandibular joint. *Journal Oral Rehabilitation 8:*293, 1981.

Robbins, S. L.: *Pathology,* 3rd ed. Philadelphia, Saunders, 1968.

Roberts, C. A.: Pers. Comm., Calvin Wells Laboratory, England.

Rockwood, C. A.: *Fractures in Children,* 4th ed. Philadelphia, Saunders, 1989.

Rockwood, C. A., and Green, D. P. (Eds.): *Fractures in Adults.* Philadelphia, Lippincott, 1975, vols 1–3.

Rogers, L. F.: *Radiology of Skeletal Trauma.* New York, Churchill Livingstone, 1982, vols 1–2.

Rong, S. H., and Zhang, Q. W.: Hydatid disease of bone. *Clinical Radiology 36:*301, 1985.

Rosenberg, N. J., Bargar, W. L., and Friedman, B.: The incidence of spondylolysis and spondylolisthesis in nonambulatory patients. *Spine 6:*35–38, 1981.

Rudolph, A. H.: Syphilis. In Hoeprich, P. D. and Jordan, M. C. (Eds.): *Infectious Diseases: A Modern Treatise of Infectious Processes,* 4th Ed. Philadelphia, Lippincott, 1989, pp. 666–684.

Ruge, D., and Wiltse, L. L.: *Spinal Disorder: Diagnosis and Treatment.* Philadelphia, Lea and Febiger, 1977.

Ryan, D. E.: Painful temporomandibular joint. In McCarty, D.J. (Ed.): *Arthritis and Allied Conditions: A Textbook of Rheumatology,* 11th ed. Philadelphia, Lea and Febiger, 1989.

Sager, P.: *Spondylosis cervicalis: a pathological and osteoarchaeological study.* Copenhagen, Munksgaard, 1969.

Saluja, P. G.: The incidence of spina bifida occulta in a historic and a modern London population. *Journal of Anatomy 158:*91, 1988.

Saluja, P. G., Fitzpatrick, F., Bruce, M. and Cross, J.: Schmorl's nodes (intervertebral ruinations of intervertebral disc tissue) in two historic British populations. *Journal of Anatomy 145:*87, 1986.

Savini, R., Martucci, E., Prosperi, P., Gusella, A., and Di Silvestre, M.: Osteoid osteoma of the spine. *Journal Ortopedic Traumatologia 14*(2):233, 1988.

Schaeffer, J. P. (Ed.): *Morris' Human Anatomy: A Complete Systematic Treatise,* 10th ed. Philadelphia, Blakiston, 1942.

Schajowicz, F.: *Tumors and Tumorlike Lesions of Bones and Joints.* New York, Springer-Verlag, 1981.

Schmidt, G. D., and Roberts, L. S.: *Foundations of Parasitology.* St. Louis, Mosby, 1977.

Schmorl, G., and Junghanns, H.: Die gesunde und kranke Wirbelsaule in Rontgenbild. *Fortschr. Rontgenstr.,* Supplement 43, Leipzig, 1932.

——: *The Human Spine in Health and Disease,* 2nd ed. (American). New York, Grune and Stratton, 1971.

Schoenecker, P. L.: Slipped capital femoral epiphysis. *Orthopaedic Review 14:*289, 1985.

Schoeninger, M. J.: *Dietary reconstruction at Chalcatzinoo, a formative period site in Morelos, Mexico.* Technical report No. 9, Museum of Anthropology. Ann Arbor, University of Michigan, 1979.

Schultz, M.: Diseases of the ear region in early and prehistoric populations. *Journal of Human Evolution 8*(6):575, 1979.

Schwarz, E.: A typical disease of the upper femoral epiphysis. In Burwell, R. B. and Harrison, M. H. M. (Eds.): *Clinical Orthopaedics and Related Research 209:*5, 1986.

Sease, Catherine: *A Conservation Manual for the Field Archaeologist.* Archaeological Research Tools. Los Angeles, University of California, 1987, vol. 4.

Seitsalo, S., Osterman, K., Poussa, M., and Laurent, L. E.: Spondylolisthesis in children under 12 years of age: long-term results of 56 patients treated conservatively or operatively. *Journal of Pediatric Orthopaedics 8:*516, 1988.

Seligsohn, R., Rippon, J. W., and Lerner, S. A.: Aspergillus terreus osteomyelitis. *Archives of Internal Medicine 137:*918, 1977.

Shaaban, M. M.: Trephination in ancient Egypt and the report of a new case from Dakleh Oasis. *Ossa 9/11:*135, 1982–1984.

Shands, A. R.: *Handbook of Orthopedic Surgery.* St. Louis, Mosby, 1951.

Sharon, R., Weinberg, H., and Husseini, N.: An unusually high incidence of homozygous MM in ankylosing spondylitis. *Journal of Bone and Joint Surgery 67*B(1):122, 1985.

Shauffer, I. A., and Collins, W. V.: The deep clavicular rhomboid fossa. *Journal of American Medical Association 195:*158, 1966.

Shipman, P., Walker, A., and Bichell, D.: *The Human Skeleton.* Cambridge, Harvard Press, 1985.

Shore, L. R.: A report of the nature of certain bony spurs arising from the dorsal arches of the thoracic vertebrae. *Journal of Anatomy 65:*379, 1931.

Sillen, A.: *Strontium and diet at Havonim Cave, Israel.* Ph.D. Dissertation. Philadelphia, University of Pennsylvania, 1981.

Simpson, W. M., and McIntosh, C. A.: Actinomycosis of the vertebrae (actinomycotic Pott's disease): report of four cases. *Archives of Surgery 14:*1166, 1927.

Sivers, J. E., and Johnson, G. K.: Diagnosis of Eagle's syndrome. *Oral Surgery Oral Medicine Oral Pathology 59:*575, 1985.

Smillie, I. S.: *Osteochondritis Dessicans.* Edinburgh, Livingstone, 1960.

———: *Injuries of the Knee Joint,* 3rd ed. Baltimore, Williams and Wilkins, 1962.

Smith, G. E.: *The Archaeological Survey of Nubia: Report on the Human Remains.* Cairo, National Printing Department, 1910.

Snow, C. E.: Early Hawaiians: *An Initial Study of Skeletal Remains from Mokapu, Oahu.* Lexington, University Press of Kentucky, 1974.

Sparks, A. K., Connor, D. H., and Neafie, R. C.: Diseases caused by cestoides. In Binford, C. H. and Connor, D. H. (Eds.): *Pathology of Tropical and Extraordinary Diseases.* Washington, D.C., Armed Forces Institute of Pathology, 1976, vol. 1.

Spjut, H. J., Dorfman, H. D., Fechner, R. E., and Ackerman, L. V.: *Tumors of Bone and Cartilage.* Washington, D. C., Armed Forces Institute of Pathology, 1971, vol 1.

Spring, D. B., Lovejoy, C. O., Bender, G. N., and Duerr, M.: The radiographic preauricular groove: its non-relationship to past parity. *American Journal of Physical Anthropology 79:*247, 1989.

Squire, L. F.: *Fundamentals of Radiology.* Cambridge, Harvard University Press, 1964.

Stafne, E. C.: Bone cavities situated near the angle of the mandible. *Journal of American Dental Association 29:*1969, 1942.

Steele, D. G., and Bramblett, C. A.: *The Anatomy and Biology of the Human Skeleton.* College Station, Texas A&M University Press, 1988.

Steinbock, T. R.: *Paleopathological Diagnosis and Interpretation.* Springfield, Thomas, 1976.

Steindler, A.: *Post-Graduate Lectures on Orthopedic Diagnosis and Indications,* Vol. III. Springfield, Thomas, 1952, vol. III.

Stewart, T. D.: The age incidence of neural-arch defects in Alaskan natives, considered from the standpoint of etiology. *Journal of Bone and Joint Surgery 35*A:937–950, 1953.

———: Examination of the possibility that certain skeletal characters predispose to defects in the lumbar neural arches. *Clinical Orthopaedics and Related Research 8:*44, 1956.

———: Distortion of the pubic symphyseal surface in females and its effect on age determination. *American Journal of Physical Anthropology 15:*9, 1957.

———: Are supra-inion depressions evidence of prophylactic trephination? *Bulletin of the History of Medicine 50:*414, 1976.

———: *Essentials of Forensic Anthropology.* Springfield, Thomas, 1979.

———: Scaphocephaly in blacks: a variant form of pathologic head deformity. *Bulletin et Memoirs de la Society d'Anthropologia de Paris, t. 9,* serie 13:267, 1982.

Stewart, T. D. and Spoehr, A.: Evidence on the paleopathology of yaws. *Bulletin of the History of Medicine 26:*538, 1952.

Stirland, A.: Pers. Comm., England.

Strassberg, M.: The epidemiology of anencephalus and spina bifida: a review. Part I: Introduction, embryology, classification and epidemiological terms. *Spina Bifida Therapy 4*(2):53, 1982.

Struthers, J.: Supra-condyloid process in man. *The Lancet, Feb:*1, 1873.

Stuart-Macadam, P. L.: Porotic hyperostosis: representative of childhood condition. *American Journal of Physical Anthropology 66:*391, 1985.

———: Porotic hyperostosis: relationship between orbital and vault lesions. *American Journal of Physical Anthropology 80:*187, 1989.

Suchey, J. M., Wiseley, D. V., Green, R. F., and Noguchi, T. T.: Analysis of dorsal pitting in the os pubis in an extensive sample of modern American females. *American Journal of Physical Anthropology 51:*517, 1979.

Suzuki, T.: Paleopathological study on osseous syphilis in skulls of the Ainu skeletal remains. *Ossa 9/11:*153, 1982–1984.

———: Paleopathological study on a case of osteosarcoma. *American Journal of Physical Anthropology 74:*309, 1987.

Swank, S. M., and Barnes, R. A.: Osteoid osteoma in a vertebral body: case report. *Spine 12*(6):602, 1987.

Sweet, P. A. S., Buonocore, M. G., and Buck, I. F.: Pre-hispanic Indian dentistry. *Dental Radiography and Photography 36:*3, 1963.

Symington, J.: On separate acromion process. *Journal of Anatomy and Physiology 34:*287, 1900.

Szypryt, E. P., Morris, D. L., and Mulholland, R. C.: Combined chemotherapy and surgery for hydatid bone disease. *Journal of Bone and Joint Surgery 69*B(1):141, 1987.

Teele, D. W., Klein, J.O., Rosner, B. A., and the Greater Boston Otitis Media Study Group: Otitis media with effusion during the first three years of life and development of speech and language. *Pediatrics 74:*282, 1984.

Terry, R. J.: A study of the supracondyloid process in the living. *American Journal of Physical Anthropology 4:*129, 1921.

———: New data on the incidence of the supracondyloid variation. *American Journal of Physical Anthropology 9:*265, 1926.

———: On the racial distribution of the supracondyloid variation. *American Journal of Physical Anthropology 14:*459, 1930.

———: Osteology. In Jackson, C. M. (Ed.): *Morris' Human Anatomy.* Philadelphia, Blakiston's, 1933.

Testut, L.: *Traité d'anatomie humaine* I. Poireir & Charphy, Paris, 1911.

Thawley, S. E., Panje, W. R., Batsakis, J. G., and Lindberg, R. D.: *Comprehensive Management of Head and Neck Tumors.* Philadelphia, Saunders, 1987, vol 2.

Thieme, F. P.: *Lumbar breakdown caused by erect posture in man: with emphasis on spondylolisthesis and herniated intervertebral discs.* Anthropological Papers, Museum of Anthropology. Ann Arbor, University of Michigan Press, No. 4, 1950.

Thomas, C. L. (Ed.): *Taber's Cyclopedic Medical Dictionary,* 15th ed. Philadelphia, Davis, 1985.

Thompson, G. H., and Salter, R. B.: Legg-Calvé-Perthes disease. *Clinical Symposia 38:*2, 1986.

Tkocz, I., and Bierring, F.: A medieval case of metastasizing carcinoma with osteosclerotic bone lesions. *American Journal of Physical Anthropology 65:*373, 1984.

Tod, P. A., and Yelland, J. D. N.: Craniostenosis. *Clinical Radiology 22:*472, 1971.

Toohey, J. S.: Skeletal presentation of congenital syphilis: case report and review of the literature. *Journal Pediatric Orthopaedics* 5(1):104, 1985.

Tosi, L.: Pers. Comm., Children's Hospital National Medical Center.

Trotter, M.: Septal apertures in the humerus of American white and Negro. *American Journal of Physical Anthropology 19:*213, 1934.

Trueta, J.: *Studies of the development and decay of the human frame.* Philadelphia, Saunders, 1968.

Twomey, L. T. and Taylor, J. R.: Age changes in lumbar vertebrae and intervertebral discs. *Clinical Orthopaedics and Related Research 224:*97, 1988.

Tyson, R., and Dyer, E. S. (Eds.): *Catalogue of the Hrdlicka Paleopathology Collection.* San Diego, San Diego Museum of Man, 1980.

Ubelaker, D. H.: *Human Skeletal Remains: Excavation, Analysis, Interpretation,* 15th ed. Washington, Taraxacum, 1989.

——: The development of American paleopathology. In Spencer, F. (Ed.): *A History of American Physical Anthropology* 1930–1980. New York, Academic, 1982.

Uemura, S., Fujishita, M., and Fuchihata, H.: Radiographic interpretation of so-called developmental defect of mandible. *Oral Surgery Oral Medicine Oral Pathology* 41(1):120, 1976.

Utsinger, P. D.: Diffuse idiopathic skeletal hyperostosis (DISH, ankylosing hyperostosis). In Moskowitz, R. W., Howell, D. S., Goldberg, V. M., and Mankin, J. J. (Eds.): *Osteoarthritis: Diagnosis and Management.* Philadelphia, Saunders, 1984.

——: Diffuse idiopathic skeletal hyperostosis. *Clinics in Rheumatic Diseases 11*(2):325, 1985.

Vailas, J. C.: Pers. Comm., George Washington University Hospital, Washington.

Verano, J.: Pers. Comm., Smithsonian Institution, Washington.

Virchow, R.: Beitrag zur Geschichte der Lues. *Dermatologische Zeitschrift 3:*1, 1896.

Waddington, M. M.: *Atlas of the Human Skull.* Vermont, Academy Books, 1981.

Walker, P. L.: Porotic hyperostosis in a marine-dependent California Indian population. *American Journal of Physical Anthropology 69:*345, 1986.

Weaver, J. K.: Bipartite patella as a cause of disability in the athlete. *American Journal of Sports Medicine 5:*137, 1977.

Walker, P. L.: Cranial injuries as evidence of violence in prehistoric southern California. *American Journal of Physical Anthropology 80:*313, 1989.

Walmsley, T.: Observations on certain structural details of the neck of the femur. *Journal Anatomy London 49:*305, 1915.

Webb, S. G.: Two possible cases of trephination from Australia. *American Journal of Physical Anthropology 75*(4):541, 1988.

Weinstein, S. L.: Natural history of congenital hip dislocation (CHD) and hip dysplasia. *Clinical Orthopaedics and Related Research 225:*62, 1988.

Welcker, H.: Cribra orbitalia. Ein ethnologish-diagnostisches merkmal am schadel mebruer menschenrassen. *Archives Anthropologie 17:*1, 1888.

Wells, C.: *Bones, Bodies, and Disease.* London, Thames and Hudson, 1964.

——: Osgood-Schlatters disease in the ninth century? *British Medical Journal 2:*623, 1968.

——: Osteochondritis dissecans in ancient British skeletal material. *Medical History 18:*365, 1974.

——: Ancient lesions of the hip joint. *Medical and Biological Illustration 26:*171, 1976.

Wells, C., and Woodhouse, N.: Paget's disease in an Anglo-Saxon. *Medical History 19:*396, 1975.

Wheat, L. J.: Histoplasmosis. In Hoeprich, P. D. and Jordan, M. C. (Eds.): *Infectious Diseases: A Modern Treatise of Infectious Processes,* 4th Ed. Philadelphia, Lippincott, 1989, pp. 481–488.

Williams, H. U.: The Origin and Antiquity of Syphilis: The Evidence from Diseased Bones. *Archives of Pathology 13:*779, 1932.

Williams, P. L. and Warwick, R. (Eds.): *Gray's Anatomy,* 36th ed. Edinburgh, Churchill Livingstone, 1980.

Willis, T. A.: Bachache from vertebral anomaly. *Surgery, Gynecology and Obstetrics May:*658, 1924.

——: The separate neural arch. *Journal of Bone and Joint Surgery 13*A:709, 1931.

Wilson, A. K.: Roentgenological findings in bilateral symmetrical thinness of the parietal bones (senile atrophy). *American Journal of Roentgenology 51:*685, 1944.

Wiltse, L. L.: The etiology of spondylolisthesis. *Journal of Bone and Joint Surgery 44*A:529, 1962.

Wiltse, L. L., Widell, E. H., Jr., and Jackson, D. W.: Fatigue fracture: the basic lesion in isthmic spondylolisthesis. *Journal of Bone and Joint Surgery 57*A:17, 1975.

Wolfe, M. S.: Pers. Comm., Washington.

Wright, V.: Osteoarthritis-epidemiology. Presented at the Conference of International Symposium on Epidemiology of Osteoarthritis. Paris, June 30, 1980.

Wyman, J.: *Observations on Crania.* Boston, A. A. Kingman, 1868.

Wynne-Davies, R.: Family studies and etiology of club foot. *Journal of Medical Genetics 2:*227–232, 1965.

Youmans, G. G. P., Paterson, P. Y., and Sommers, H. M.: *The Biologic and Clinical Basis of Infectious Disease.* Philadelphia, Saunders, 1980.

Zaaijer, T.: Untersuchungen uber die form des beckens javanischen Frauen, Naturrk. Verhandel. Holland Maatsch. *Wentesch Haarlem 24:*1, 1866.

Zaino, E.: *Symmetrical osteoporosis, a sign of severe anemia in the prehistoric Pueblo Indians of the Southwest.* In Wade, W. (Ed.): Miscellaneous Papers in Paleopathology. Museum of Northern Arizona Technical Series No. 7, 1967.

Zimmerman, M., and Kelley, M.: *Atlas of Human Paleopathology.* New York, Praeger, 1982.

AUTHOR INDEX

A

Ackerman, L. V., 34, 35, 112, 113, 114, 184
Adams, J. L., 87, 120, 167
Afanasiev, R., 53, 177
Ahmed, A., 169
Akrawi, F., 8, 167
Ala-Ketola, L., 20, 181
Albright, F., 49, 167
Alexander, W. J., 76, 178
Allbrook, D. B., 48, 167
Allen, H., 96, 167
Allman, R. M., 106, 168
Altchek, M., 127, 167
Amaral, B. W., 141, 169
Anderson, J. E., 74, 131, 167
Anderson, T., 167
Angel, J. Larry, 8, 9, 21, 22, 29, 34, 83
Arensburg, B., 63, 171
Arey, 15
Aria, A. A., 31, 172
Armelagos, G. J., 21, 22, 23, 24, 146, 167, 169, 172, 176, 178
Atlah, C. A., 141, 167

B

Bacon, P. A., 4, 19, 20, 47, 48, 50, 55, 58, 61, 72, 88, 93, 112, 120, 129, 171
Baker, B. J., 22, 146, 167
Baker, D., 53, 172
Bamji, A. N., 4, 19, 20, 47, 48, 50, 55, 58, 61, 72, 88, 93, 112, 120, 129, 171
Bargar, W. L., 53, 182
Barnard, L. B., 87, 167
Barnes, R. A., 60, 185
Barranger, J. A., 132, 171
Barrie, H. J., 88, 102, 104, 167
Barrow, M. V., 128, 180

Barry, H. C., 34, 167
Bass, W. M., xiii, 8, 11, 87, 167, 178
Bassiouni, M., 129, 168
Batsakis, J. G., 30, 185
Bazergui, A., 53, 176
Bender, G. N., 76, 79, 184
Bennett, P. H., 56, 173
Benson, D. R., 63, 180
Bergman, 71, 87
Berry, A. C., 31, 168
Berry, R. J., 31, 168
Bertaux, T. A., 96, 168
Bichell, D., 11, 183
Bierring, F., 148, 186
Binford, C. H., 141, 143, 144, 154, 168, 184
Bingle, G. J., 128, 180
Blackburne, J. S., 53, 168
Blackwood, H. J. J., 32, 168
Bloch, D. A., 20, 176
Bloch, L., 8, 168
Bloom, R. A., 53, 172, 177
Bloomberg, 68
Bluestone, C. D., 27, 168
Bowen, J. C., 98, 168
Bowen, J. R., 98, 168
Bradford, D. S., 48, 50, 53, 62, 63, 168, 179
Bradley, J., 102, 168
Bradtmiller, Bruce, xiii, 11
Brahney, C., 31, 172
Bramblett, C. A., 11, 73, 184
Bridges, P. S., 53, 168
Bried, C. J., 141, 168
Brook, C. J., 141, 168
Brothwell, D. R., 4, 7, 8, 11, 148, 168, 173, 177, 179
Brower, A. C., 85, 106, 138, 168
Brown, T., 42, 182
Bruce, M., 52, 183
Brunker, L., 174

189

Buck, I. F., 25, 185
Buckley, J. H., 55, 172
Buikstra, J. E., 8, 170
Bullough, P. G., 54, 178
Bulos, S., 52, 169
Bunnell, W. P., 52, 169
Buonocore, M. G., 25, 185
Burke, M. J., 20, 168
Burman, M. S., 126, 169
Burston, W. R., 23, 177
Burwell, R. B., 168, 183
Burwell, R. G., 97, 175
Busch, M. T., 98, 169
Bushan, B., 32, 169

C

Cady, 87
Caffey, J., 85, 137, 169
Cain, P. T., 141, 169
Caldwell, J. R., 20, 181
Calin, A., 58, 169
Callahan, 120
Campbell, J. P. S., 30, 176
Campillo, D., 52, 169
Canty, A. A., 27, 171
Caraveo, J., 141, 169
Carlson, D., 22, 23, 169
Carney, C. N., 65, 113, 118, 169
Casscells, S. W., 105, 169
Caterall, A., 98, 169
Cave, A. J. E., 70, 81, 169
Cerruti, M. M., 141, 167
Chopra, S. R. K., 23, 169
Chou, S. N., 168
Chung, S. M. K., 83, 169
Claffey, T. J., 93, 169
Clanton, T. O., 88, 102, 169
Coates, C. J., 91, 178
Cochran, R. M., 127, 175
Cochrane, R. G., 181
Cockburn, A., 4, 8, 169
Cockburn, E., 4, 8, 169
Cockburn, T. A., 169
Cockshott, P. W., 82, 169
Cohen, M. M., Jr., 27, 170
Coley, B., 148, 170
Colles, Abraham, 91
Collins, W. V., 81, 183

Congdon, R. T., 53, 170
Conner, D. H., 141, 143, 144, 154, 168, 184
Cook, D. C., 8, 170
Correll, R. W., 30, 31, 170
Corruccini, R. S., 25, 173
Courtenay, B. G., 102, 175
Cowell, H. R., 51, 170
Cowell, M. J., 51, 170
Crelin, E. S., 20, 170
Cross, J., 52, 183
Cummings, S. R., 49, 98, 170
Cushing, A. H., 141, 171

D

Dahlin, D., 114, 148, 170
Dalinka, M. K., 141, 170
Dallman, P., 22, 170
Dandy, D. J., 102, 168
Daniels, 109
Dannels, E. G., 170
Danzig, L. A., 85, 170
Daveny, J. K., 141, 170
Davey, T. F., 181
David, D. J., 27, 37, 170
David, R., 35, 170
Davis, P. R., 48, 60, 170
Day, S. B., 170
De Chazel, 30
De Lee, J. C., 88, 102, 169
Derry, D. E., 72, 170
Dhem, A., 23, 178
Dhooria, H. S., 32, 171
Dickel, D. N., 50, 72, 170
Dickson, R. A., 62, 171
Dieppe, P. A., 4, 19, 20, 47, 50, 55, 58, 61, 72, 88, 91, 93, 112, 120, 129, 171
Dinari, G., 53, 177
Dinbar, J. S., 178
Dinnenberg, S., 141, 170
Di Silvestre, M., 60, 183
Doran, G. H., 50, 72, 170
Dorfman, H. D., 34, 35, 112, 113, 114, 184
Douglas, 30
Duerr, M., 76, 79, 184
Dugdale, A. E., 27, 171
Duncan, M. H., 141, 171
Dutour, O., 93, 94, 171

SUBJECT INDEX

199